YouTube
An Insider's Guide to Climbing the Charts

Alan Lastufka and Michael W. Dean

O'REILLY®

Beijing · Cambridge · Farnham · Köln · Sebastopol · Taipei · Tokyo

YouTube: An Insider's Guide to Climbing the Charts
by Alan Lastufka and Michael W. Dean

Copyright © 2009 Alan Lastufka and Michael W. Dean. All rights reserved.
Printed in the United States of America

Published by O'Reilly Media, Inc. 1005 Gravenstein Highway North, Sebastopol CA 95472

O'Reilly books may be purchased for educational, business, or sales promotional use. Online editions are also available for most titles (*safari.oreilly.com*). For more information, contact our corporate/institutional sales department: (800) 998-9938 or *corporate@oreilly.com*.

Editor: Sandy Doell	**Compositor:** Ron Bilodeau
Production Editor: Michele Filshie	**Indexer:** Ted Laux
Technical Editor: Chris Caulder	**Illustrator:** Jessamyn Read
Copyeditor: Kim Wimpsett	**Interior Designer:** David Futato
Proofreader: Nancy Bell	**Cover Designer:** Monica Kamsvaag

Print History:

November 2008:   First Edition.

RepKover.
This book uses Repkover,™ a durable and flexible lay-flat binding.

ISBN: 978-0-596-52114-1
[M]   [1/09]

This book is dedicated to my family:
Carol and John, Steve and Pat, Mark and Terra, for all your love and support
and to my best friends Danny, Aaron, and Todd.
Thanks for your unwavering ability to put up with me, and encourage me.

—Alan Lastufka

Contents

Foreword

Let me begin by making an observation that the authors are too humble to make for themselves: This book will change your life. But more on that in a moment. First I'm going to do something prototypically YouTubian: I'm going to talk about myself.

Like most people, I got to know YouTube as part of my lifelong project to discover every method of procrastination available on the Internet. I'd go online and watch music videos or laughing babies or squirrels on water skis or whatever the new viral video happened to be. But I didn't get it, not really. I didn't understand the immense potential of YouTube, because what is truly powerful about the site is its community-building features.

In the summer of 2006, I discovered two very different YouTube channels and fell in love. The first was askaninja, a hilarious show in which a man dressed as a ninja answers viewers' questions. The quick-cut editing and rapid-fire speech of askaninja have since become a hallmark of countless popular videos (including mine). The second was lonelygirl15, a seemingly real vlog made by a seemingly real 16-year-old girl whose parents were caught up in a profoundly weird occultist religion. *Lg15*, as the show was known, felt to me like all the best parts of *Lost*, but it was even better. It felt *real* in a way that no television show ever could.

After watching it for a few months, I began to understand why I enjoyed it so much: I was participating in the creation of the show. My video responses (Chapter 8) appeared on the *lg15* YouTube page, as did my text comments. I interacted with other fans and became friends with them. Television shows don't bring together strangers like YouTube communities can, because a TV show—even an awesome one—is merely something you watch. TV does not get more awesome as a result of you watching it, but YouTube channels do. That's the miracle of YouTube: You get to help make the *awesome*.

And that's where the changing-your-life part comes in. In this book, you'll learn the technical details of how to create and edit a video, how to maximize its quality for the Web, and how to upload it to YouTube. These are vital skills for anyone seeking to make web-based video content. But you'll also learn how to find an audience on YouTube—how to get your content to the millions of people who watch and enjoy online videos. YouTube is so huge and there are so many videos that it can at times feel impossible to get anyone to watch yours. Building a base of subscribers and fans is tremendously challenging, but no one in the world knows more about how it works than YouTube star Alan Lastufka. Along with coauthor Michael W. Dean (a well-known independent filmmaker and author), Alan will show you a variety of effective strategies for becoming more popular.

My brother and I made vlogs back and forth to each other every day for seven months before any of our videos received 1,000 views. Then our audience started to grow, and with Alan's help, it has continued to grow ever since. We have more than 40,000 subscribers now—a vibrant community of smart, thoughtful people celebrating nerd culture and brotherhood. The people of the YouTube community have changed my life for the better, and in some small way, I've changed theirs, too. Not every channel will have a million viewers, of course, but every channel can participate in the growth of exciting and passionate online communities.

I don't know of a more vibrant creative sandbox than YouTube. There are already so many great writers, artists, singers, comedians, filmmakers, and actors making You-Tube great. But by the very design of the site, there is always room for more—for a new perspective or a niche we didn't even know existed.

So, let's get started! Proceed into the wisdom of this book, and I look forward to seeing you pass me on the Most Subscribed list. Best wishes!

—John Green (vlogbrothers)

John Green is one half of the YouTube channel vlogbrothers, along with his brother Hank (Hank is interviewed in Chapter 15).

www.youtube.com/user/vlogbrothers (URL F.1)

John is *The New York Times* bestselling author of *Paper Towns*. His novel *Looking for Alaska* won the American Library Association's 2006 Michael L. Printz Award. He also wrote *An Abundance of Katherines* and has been covered in the *Wall Street Journal* and on CBS News. His personal site is *www.sparksflyup.com*.

Preface

Why You Should Buy This Book

Greetings, eager YouTuber!

I'm Michael W. Dean, one of your co-authors (along with Alan "fallofautumndistro" Lastufka), here to introduce you to the book. This part of a how-to book is usually called "Who This Book Is For." It usually contains a description of the technology being discussed and why you should "get in now on the ground floor."

We're not going to do that. The ground floor of YouTube is well in the past. This is not an emerging technology, even though it's been active for only a few years. YouTube launched in 2005, which makes it, in Internet terms, long established. The 'Tube is such a massive cultural force that you already know what it is and how important it is. That's why you're in the bookstore looking at a book about YouTube.

This section of a how-to book usually paints a picture of how much money you can make, and how you might get famous, in words written by someone who is neither. We're not going to do that.

This book is for people who want to have visibility on YouTube and don't like being lied to. A lot of books lie.

There are other books out there about YouTube but you should buy this one. Seriously. Here's why. This book doesn't promise that you'll be famous or rich. Any book that does is lying to you, and you should avoid books that lie to you.

If you like being lied to, grab one of the books on either side of this one on the same shelf in the bookstore. Those books will be happy to take your money, promise you the world, and lie to you in the process.

What This Book Promises

If you're smart (and you *are* smart, because you picked this book), talented (and you *are* talented, because this book attracted you and this book is about talent and vision), and have some good ideas, you'll make quality art and have a damn good chance of getting your work seen. Other than that, it's up to the universe. Viral marketing can't be created on demand, at least not at a corporate level, although many professional consultants are willing to lie to you, take your money, and say it can be done. There's no guarantee since computer voodoo and chance are not sciences. But viral marketing is something that can be given a good chance of happening. It can be "incubated".

This book will, however, show you everything you need to know to have a good chance. That's more than the other YouTube books will give you. We've read them all and use most of them to hold down piles of paper near windows and hold up wobbly legs on old tables.

> **Note** *Viral*, as in *viral marketing* and *viral videos*, in the context of this book has nothing to do with computer viruses. It means "videos that are passed from one excited person to another via email and YouTube—by Internet "word of mouth"—and spread like a thought virus. Alan has had a lot of videos "go viral." You want your videos to go viral. The main goal of this book is to teach you everything we know about going viral. You want to go viral.

Most of those other YouTube books are written by people who don't have many (or any) subscribers on YouTube. They're written by either marketing weenies, college dorks with mostly theoretical knowledge, or tech writers who were assigned the project, then signed up for a YouTube account, and tried to get "up to speed" so they could write the book.

My writing partner on this book, my good friend Alan "fallofautumndistro" Lastufka, is a rock star on YouTube:

www.youtube.com/fallofautumndistro (URL P.1)

Alan may not be in the top 5 percent on YouTube and he's never been featured on the Leno show or parodied on *South Park*, but he's pretty well known. He has many thousands of subscribers, makes money on YouTube, and has been featured many times.

Alan "fallofautumndistro" Lastufka.

(*Featured* means spotlighted by the people who run YouTube.) Alan's had features on the front page of YouTube.com, and most of his videos get many thousands of views. Some get upwards of a million views. He was even invited to write and direct Lisa Nova in one of his past YouTube videos. (LisaNova is huge on YouTube, was a regular on MADtv, and got that gig through her YouTube work.) Alan knows what he's talking about, and you should listen to him.

Alan also has a long history in the punk scene, which totally transfers to YouTube. The type of drive-by D.I.Y. (do-it-yourself) marketing and distribution that Alan did with his CD/zine/book/video distro company, and currently does on his large indie T-shirt printing company, are like grad school for social networking environments like YouTube. It's experience you can't get in college. And there are currently some very successful YouTubers who are in the top 5 percent offering Alan a good salary to leave his cold, small town in Illinois to move out to Southern California, be set up with a place to live, lie on the beach, and do their YouTube marketing from his laptop. That is not the case with the folks who wrote those other YouTube books.

Alan is very active in the YouTube community. The community is mentioned only in passing in the other books, when it's mentioned at all, because the people writing those books are outsiders. Alan is all about the community and so is this book. Community is the key to YouTube. It's how you go viral.

> **Note** Alan is calm, driven, and meticulous, but still damn lively. I'm hyper and in your face. The mixture works because we're both detail-oriented, meticulous, efficient, and professional. And we'll teach you how to be all of this yourself. There's no other tech book like this. This is less like a textbook and more like two friends hanging out with you and explaining things. We think you'll enjoy the process.

I have fewer YouTube subscribers than Alan: *www.youtube.com/user/Kittyfeet69* (URL P.2) but still more than any of the other people who've written books on the subject so far. (I've checked their YouTube pages, at least the ones who even have YouTube pages.) I, too, have been featured by the editors of YouTube, and some of my videos have gotten thirty thousand views. I've also toured the world in a punk band named Bomb: *www.hitsofacid.com* (URL P.3)

Bomb was on a major label (Warner/Reprise) and on several indie labels. I have decades of D.I.Y. marketing experience from that, and it transfers to doing things on YouTube. I also have a serious background in filmmaking. I've made two feature documentary films: *Hubert Selby Jr: It/ll Be Better Tomorrow* (narrated by Robert Downey Jr.) and *D.I.Y. or DIE: How to Survive as an Independent Artist*: *www.blockbuster.com/catalog/movieDetails/217632* (URL P.4)

Both films have distribution and make money from DVD sales. I've traveled the U.S. and Europe showing them and doing Q&A in person. *Hubert Selby Jr: It/ll Be Better Tomorrow* was favorably reviewed in the high-end Hollywood trade paper *Variety*. I was flown to France to premiere the film in a major film festival: *www.variety.com/review/VE1117928194?categoryid=31&cs=1* (URL P.5)

Most of the people writing YouTube books haven't made a feature film. I wrote the book *$30 Film School*: *www.amazon.com/gp/product/1598631896/* (URL P.6), published by Course Technology, 2002, 2008.

You can go buy one of those other YouTube books, if you'd rather be told fun things like "You'll be famous, get rich, and find a mate on YouTube if you buy this book." We don't guarantee any of that. (But Alan met one of his previous girlfriends on YouTube, and she's really smart, funny, and beautiful.) *$30 Film School* is taught in many colleges. So, you should probably listen to me too.

If you want to hear the truth, set realistic (yet very cool) goals and hang out with two guys who know what you need to know, sit back, grab your favorite beverage, kick up your feet, and absorb the wisdom of the 'Tube, from two cats who live it.

Here's a picture of me.

Michael W. Dean.

How This Book Is Organized

Chapter 1, "What Is This YouTube of Which You Speak?," includes a history of You-Tube, explaining how it came into being and why it found such an eager following. Michael explains the joys and problems of the "level playing field" that is YouTube. Alan gives information on how to navigate YouTube and what it means to *go viral*.

In Chapter 2, "Storytelling and Directing," and Chapter 3, "99-Cent Film School: Shooting, Editing, and Rendering," Michael tells you about making quality videos that people will watch and pass on to friends. Alan shows you how to edit and render the videos, so you can upload them to YouTube.

Chapter 4, "Creating Your Very Own Channel," is where Alan walks you step-by-step through the process of registering your channel, and "pimping your profile," so that your spot on YouTube will stand out from the crowd.

In Chapter 5, "Broadcasting Yourself: User Generated Content," Alan walks you through putting your well-constructed videos out there for the rest of the world to see. He also provides insight into the more common "daily use" aspects of YouTube, such as creating relevant thumbnails, *favoriting*, creating a playlist, flagging, and more.

In Chapter 6, "Rebroadcasting Commercial Content," Michael guides you through the tricky landscape of using other people's work in your videos. You'll learn about copyrights, fair use and parody, public domain, Creative Commons and lots more.

Chapter 7, "Building Your Audience," is where Alan tells you all you'll need to know about growing in popularity, making friends, and becoming well known in the higher social stratosphere of YouTube.

In Chapter 8, "The Community: Where Do You Fit In?" Alan talks about his experiences making collaboration videos, networking, collaboration channels, and some of his own collaborations, including *fiveawesomeguys*, a top group on YouTube.

In Chapter 9, "Hacking the System: How to Cheat (and Why You Shouldn't)," Alan shares what he's learned about nefarious dealings on YouTube—what's considered acceptable behavior and what can get you kicked out for good (and how to get back on if you're kicked off). Michael gives a fascinating history of hacking in general.

Chapter 10, "Reaching the World," is where Alan tells you all about the social networking world of YouTube, blogging, vlogging, signatures, social bookmarking, and just generally becoming a pro at making friends.

In Chapter 11, "Money Money Money!" Alan tells you about the Partner Program on YouTube, how to apply, and how to finally *monetize* your work.

Chapter 12 is called "Beyond the 'Tube," and here you'll learn from Alan and Michael how to create your own identity, independent of YouTube. We also include Michael's tricks of time budgeting to still have a life away from the computer.

Chapter 13, "Becoming a Success Story," is about becoming a star in your own right, getting featured, not just on YouTube's front page, but on the nightly news, turning up first in Google searches, and generally becoming well known.

In Chapter 14, "Closing Arguments," Michael philosophizes about what it means to be a true Renaissance artist in the Web 2.0 realm and shares real gems of wisdom.

Chapter 15, "Interviews With Other YouTube Rock Stars," finds Alan interviewing some of the huge stars of YouTube.

Bonus Materials Online

Visit and subscribe to *www.viralvideowannabe.com* (URL P.7) for ongoing updates, tips, and techniques about YouTube. We've also included two bonus chapters exclusively on the website, including:

Bonus Chapter 1, "Recommended Reading, Surfing, and Viewing": In this chapter, Alan recommends some of the better YouTube channels and videos: *http://viralvideowannabe.com/bonus-chapter-01* (URL P.8)

Bonus Chapter 2, "Gear Manufacturers' Websites, Addresses, and Contact Information": Michael and Alan list some of the equipment and software they use and recommend. If you haven't yet purchased your camera or editing software, this would be a great place to start your research.

http://viralvideowannabe.com/bonus-chapter-02 (URL P.9)

RSS sidebar: Michael tells you how to create a subscription-based feed for your videos, so people can take you with them on their iPods and watch your short form masterpieces away from their computer. You can't do that with YouTube!

http://tinyurl.com/6pfuls (URL P.10)

How to Use URLs in This Book

This book has a lot of URLs (web addresses), specifically a lot of YouTube addresses. YouTube addresses are made from random letters, like this:

www.youtube.com/watch?v=vbZdcEoXeGQ (URL P.11)

Those are kinda hard to type correctly, so there's a lot of room for error. They're made for linking, not for typing. So, we've taken care of that for you by putting a link number (formatted as a *chapter.link*, as in *URL 1.1*, *URL 1.2*, *URL 1.3* for the first three links in Chapter 1) for every link in this book, and then we've linked them for you on a page on Alan's site, here:

http://viralvideowannabe.com/ytlinks (URL P.12)

Type that in your browser once, bookmark it, and you can easily reference every link to videos and websites in this book without having to actually type them.

The URLs are mirrored on the O'Reilly site, too:

www.oreilly.com/catalog/9780596521141 (URL P.13)

On the formatting of URLs

There are tradeoffs when making art with other people. Sometimes I write books and self-publish them. I have complete creative control but don't reach as many people or make as much money as when I write books for a company. When doing art with a company, it's a balancing act with other people's ideas. I try to fit each project to the right method of distribution, and it seemed to me that this book was better suited to being done with a company, so I had to be a team player. I can wear that hat.

I got along great with the team working on this book. The only disagreement we had was about the formatting of URLs. I contended that when a sentence ends with a URL, there should not be a period at the end of the URL. I was voted down on this. So, in this book, if a URL falls at the end of a sentence, it will be followed by a period. I think this is confusing, and even though it does conform to standard rules of English, I believe we're at a time in progress where that rule should change, because URLs with periods at the end don't work. Most people who have been on the Internet for a while know not to type the period, or a dash added by a line break. We've provided a page with clickable links without the periods but I had to state my feelings on this, because there are probably a few people new to the Internet reading this book who would type the URL with the period and wonder why the URL didn't work if I didn't explain it.

Conventions Used in This Book

Notes indicate an expansion on the surrounding text. Pro Tips are short sidebars.

> **Note** This is a note. It contains useful information about the topic at hand, often highlighting important concepts or best practices.

PRO TIP

Indicates a piece of advice that could only be provided by someone who is a You-Tube professional. Alan and Michael are both YouTube professionals; that is, they have been paid for their artistry, either on YouTube or in the making of film, music, books, and videos. A pro tip is a tip from someone who knows how to make money on YouTube.

Safari® Books Online

Safari Books Online   When you see a Safari Enabled icon on the cover of your favorite technology book, that means the book is available online through the O'Reilly Network Safari Bookshelf.

Safari offers a solution that's better than e-books. It's a virtual library that lets you easily search thousands of top tech books, cut and paste code samples, download chapters, and find quick answers when you need the most accurate, current information. Try it for free at *http://safari.oreilly.com*. (URL P. 14)

How to Contact Us

Please address comments and questions concerning this book to the publisher:

O'Reilly Media, Inc.
1005 Gravenstein Highway North
Sebastopol, CA 95472
800-998-9938 (in the United States or Canada)
707-829-0515 (international or local)
707-829-0104 (fax)

Access the web page with errata, examples, and any additional information here:

www.oreilly.com/catalog/9780596521141 (URL P.15)

To comment or ask technical questions about this book, send email to this address:

bookquestions@oreilly.com (URL P.16)

To learn about our books, conferences, the O'Reilly Network and more, go to:

www.oreilly.com (URL P.17)

Acknowledgments

Alan Lastufka: I would like to acknowledge the people who make YouTube worth it: Charlie McDonnell, Alex Day, Todd Williams, and Johnny Durham for sharing their lives with me on the fiveawesomeguys channel; Lisa Donovan, Michael Buckley, Hank and John Green, Kevin Nalty, Liam Kyle Sullivan, and George Strompolos for speaking within these pages; and everyone else I've had the pleasure of collaborating and making art with. Thanks to Lea, achampag, for your friendship. Lesliefoundhergrail for cheerleading me through this. Thanks also to Liron Steinfeld of blogTV for her assistance and support.

Michael W. Dean: Sweet kisses for Debra Jean Dean. She's my best friend, wife, and partner in life, art, cat wrangling, and other niftiness. Thank you for reading all the drafts as well as for making life the eternally giddy nesting experience I always knew it should be.

Alan and Michael: We'd like to thank our first-rate team on this book: Steve Weiss, Michele Filshie, and Dennis Fitzgerald at O'Reilly for making our anarchy work; and Sandy Doell and Chris Caulder for expertly dotting the p's and crossing the q's. Thanks also to Kim Wimpsett, Nancy Bell, and Ted Laux.

About the Authors

Alan Lastufka a.k.a. fallofautumndistro (YouTube user name), is one of the Top 100 Most Subscribed Comedians on YouTube. His YouTube videos have had more than three million total views. One of his early short films received multiple airings on the Independent Film Channel.

On YouTube, Lastufka is widely praised for his collaboration videos. One of the most popular is the "iPwn" iPhone parody commercial, starring MadTV cast member LisaNova (as well as some narration at the end that was engineered and produced by Michael Dean).

Alan is currently a teacher and skills coach for a non-profit social services agency.

Michael W. Dean is a filmmaker/author/musician/podcaster who directed the films *D.I.Y. or DIE: How To Survive as an Independent Artist* and *Hubert Selby Jr: It/ll Be Better Tomorrow* and wrote the book *$30 Film School*. He has toured the US and Europe extensively, lecturing at youth centers, colleges, and museums on the subject of D.I.Y. (do-it-yourself) art production and promotion.

Dean has written for *MAKE Magazine*, writes for the O'Reilly Digital Media site and runs the pop culture blog, StinkFight.com. He was a contributor to the book *Digital Video Hacks*, and edited *DV Filmmaking: From Start to Finish* (both from O'Reilly).

Colophon

The cover fonts include Stamp and Sketchathon. The text font is Minion Pro and the heading font is Adobe Myriad Pro Condensed.

What Is This YouTube of Which You Speak?

Alan Lastufka

YouTube and Online Video: A Brief History

Before sites like YouTube made sharing video online an everyday occurrence, it was very difficult to show your friends and family your latest clip. Weird errors having something to do with "missing codecs" were usually what you found staring back at you from your CRT monitor, rather than your nephew's first steps or your brother's extreme bike stunts. YouTube founders Chad Hurley and Steve Chen were tired of this happening to them, and they assumed others were too.

Together with Jawed Karim, the three former PayPal employees created a video-sharing service that would be accessible to almost every single computer with an Internet connection. You could upload all varieties of file formats and compression settings, and YouTube's behind-the-scenes magic (server computers and proprietary software) would process the file and give you a link to your video that could easily be emailed to anyone and viewed on any connection from dial-up to a T1 line.

> **Note** It's okay if you don't understand terms like *compression settings* or *codecs* right now; we will explain these, and other technical terms, as they pertain to your videos later in this book.

Although in hindsight it may seem like a no-brainer, YouTube was the first of its kind. The potential for millions of people, from experienced Hollywood filmmakers (which YouTube has) to 16 year olds who just received their first webcams (which YouTube also has) could post their work in the same arena, with the same tools available to both from the same website.

Jawed uploaded the very first video to YouTube, a short clip of himself at the zoo, in April 2005, *www.youtube.com/watch?v=jNQXAC9IVRw* (URL 1.1).

The new YouTube site didn't instantly take off. In the early days, YouTube tried a number of promotional gimmicks to attract users, including offering money via ads on Craigslist to hot, young women who would post videos on the site, as well as giving away free iPods to random active users every day for a month.

Eventually, these gimmicks, in addition to word of mouth and quite a bit of press, garnered some solid traffic for YouTube. But in the beginning, there wasn't anything to keep someone for more than a few minutes; they watched one or two clips and then clicked elsewhere. This changed very rapidly, however, as a community of users formed. This community inspired, supported, and rivaled each other—something the founders never anticipated. Chad and Steve assumed most users would utilize the site to send family video to out-of-town relatives. But users started registering and uploading videos of themselves simply talking about their lives and their opinions and asking others about theirs. These users came to be known as *vloggers* (video bloggers), and these vloggers quickly formed a community. (*Blog* means web log, a frequently updated online text diary of sorts.)

Note Vloggers existed before YouTube and have even found fame outside of YouTube; the popular daily vlogger Ze Frank (who is now retired) is one such example. Vlogging is simply sharing your life, your opinions, and your feelings with others via video. The term is not exclusive to YouTube.

The community has, in fact, become such a powerful force on YouTube that we've devoted all of Chapter 8 to it, and we will mention the community often throughout the other chapters.

As the number of YouTube visitors continued to grow, so did the press. *Time Magazine* named "You" the Person of the Year for 2006 and helped cement YouTube as the number-one site for sharing videos online. So large is its following that YouTube streams more videos each day than all of its major competitors, including MySpace and AOL Video, combined.

Quickly taking note of YouTube's unique user base and functionality, Google, the search engine giant, purchased YouTube for more than $1.6 billion dollars in 2006. Google could offer YouTube users more server bandwidth (faster service) and better promotional tools (the integration of AdSense for YouTube Partners, covered in Chapter 11), along with Google's patented, powerful search engine.

Finding Videos

Before you get caught up in making your videos, it would be wise to watch what has come before you. Just as no great artist emerges without a few drawing classes, I believe it's important for you to be exposed to previous videos that have worked and a few that haven't. This book is full of links to wonderful videos, and if you've already bookmarked the links page from the preface, *www.viralvideowannabe.com/ytlinks* (URL 1.2), feel free to click any of them before finding them in the book. You may want to take a moment, put this book down, and watch a few of those videos.

If, instead, you're looking for a video on a specific subject, the best place to start is YouTube's search function (see Figure 1-1).

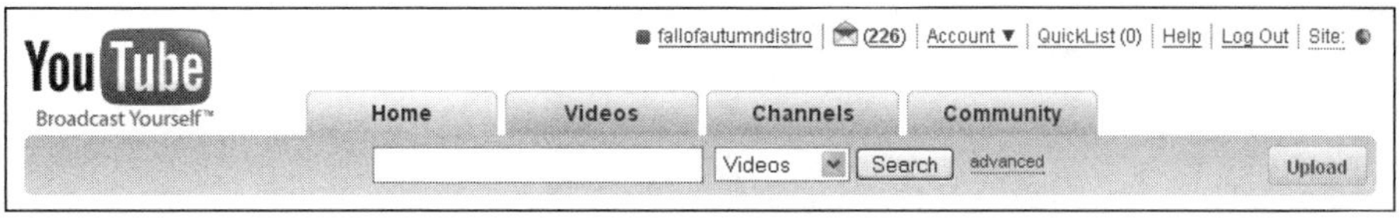

Figure 1-1. At the top of each page on YouTube, you will find the search function.

Near the top of almost every page on YouTube, you will find the search function. From here, you can search for videos or channels. For general searches, this works great. But if you want to refine your search (perhaps you're looking for a certain video you saw last week but can't remember the name or who made it), there are the advanced search functions. Simply click the word *advanced* to the right of the Search button, and the options shown in Figure 1-2 appear.

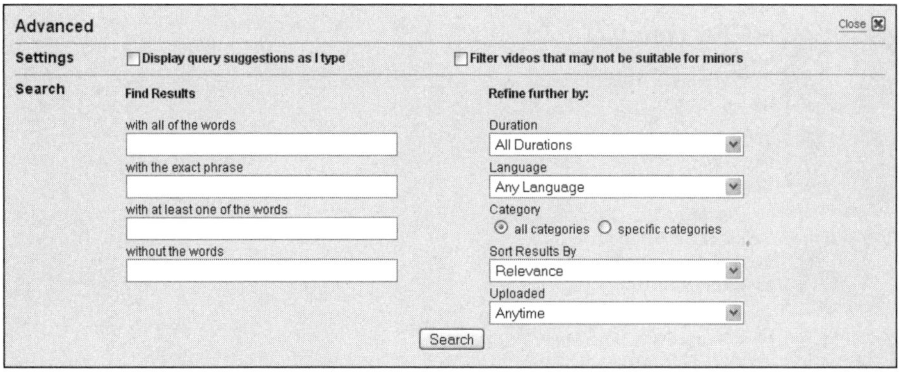

Figure 1-2. The advanced search functions.

The advanced search functions allow you to search using numerous criteria to help refine your search. YouTube has millions of videos, and it's very likely that even a specific search could return dozens of videos that match your search criteria. Use the advanced features as often as you can.

Another popular, though less reliable, way to find videos is the Related Videos feature on YouTube. After each video you watch, a list of eight related videos appears, rotating two videos at a time (see Figure 1-3 and Figure 1-4).

Figure 1-3. Related videos show up in the player after a video finishes.

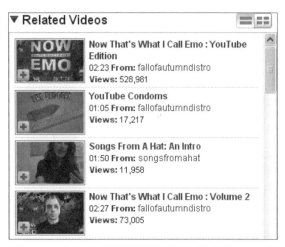

Figure 1-4. Related videos also display to the right of the player on each video page.

Google has a history of keeping its search algorithms (the processes it uses to choose search results, or, in this case, related videos) a closely guarded secret, so there is no way to say with certainty how the related videos are chosen. I've watched this process in operation over a few weeks, and it appears that when you create a video and upload it on YouTube, related videos are determined based on the tags and titles you assigned to it. Later, the process updates, and new related videos appear, based on what users watched just before or just after they watched your video.

The related videos, as with most of YouTube, seem to be a level playing field, meaning your video with 20 views could easily place right beside the latest viral video with 200,000 views. It's this level playing field that keeps new YouTubers hopeful; without it, the same channels would receive promotion time and time again, and the content would quickly become stale. Later in this chapter, Michael will discuss some of the pros and cons of this level playing field.

PRO TIP

If you'd like to increase the chance that your video will turn up in the Related Videos section, be sure to use similar tags and video description. You can also leave your video as a response to the video you'd like yours to relate to. Viewers tend to click on video responses, and this will help YouTube relate your video to the one in question. We'll cover how to leave Video Responses in Chapter 8.

Navigating YouTube

Okay, so you searched for your favorite band's latest music video, and you searched your name just to see what comes up, but all you've been watching are old videos that sit at the top of the search results—where's all the new stuff?

It's on the lists. YouTube keeps numerous lists, defaulted to the top videos of the day (on YouTube, a day technically lasts 48 hours, but we'll talk more about that in Chapter 13). Included among these lists are the Most Viewed videos of the day, which is as simple as it sounds; they are the top 100 videos that have received the most views (see Figure 1-5). Other lists include Most Discussed, which lists the top 100 videos with the most text comments, and Top Rated, which lists videos with the most five-star ratings.

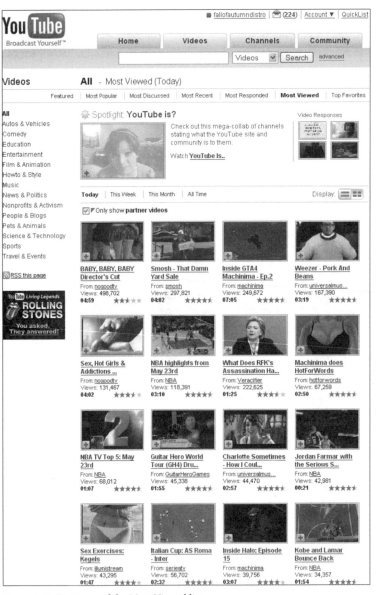

Figure 1-5. First page of the Most Viewed list.

To find these lists, simply scroll up to the top of any YouTube page, and click the Videos tab, as shown in Figure 1-6. Clicking the Videos tab takes you first to the Most Viewed videos of the day. From there, you can select from the various lists, as mentioned earlier, to find which videos people are currently viewing the most, which are being discussed the most, and so on.

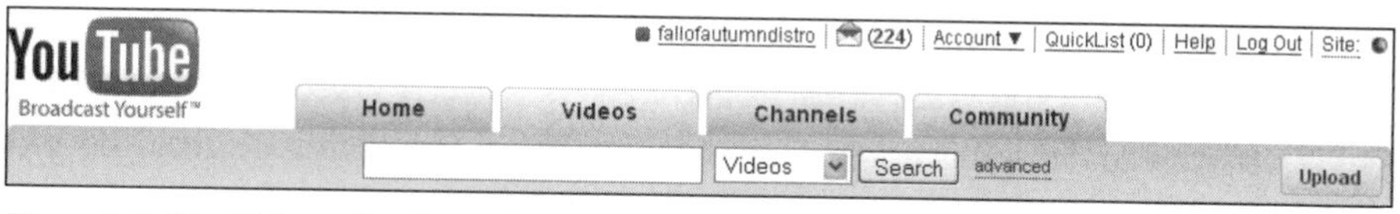

Figure 1-6. The Videos tab is located at the top of each page.

Again, these lists do not favor one type of account over another. It may be harder for someone with few subscribers to gain the views quick enough to land on the Most Viewed list for today, but it does happen when the right video is seen by, and shared with, the right people. This is called *going viral*.

Going Viral

As Michael stated in the preface, you want to go viral. A viral video is one that is so funny, so outrageous, or so shocking that it is immediately shared from one viewer to the next, and on and on; it goes viral, like the flu. (But in a good way.) We've all received emails or instant messages containing links to funny parody songs, laughing babies with gas, or shocking accidents caught on tape and then uploaded to YouTube. Most viral videos used to happen by accident. They were not planned in advance or scripted or even framed properly, because they were happening in the moment.

These days, however, the majority of viral videos are well-produced, scripted, and sometimes even funded productions. Larger comedy groups and corporations saw the power in numbers of early viral videos and have spent a lot of time, and money, trying to reproduce those organic viewing frenzies time and time again.

THE FIRST VIRAL VIDEOS

Gary Brolsma is credited with one of the first viral videos, *Numa Numa*.

www.youtube.com/watch?v=60og9gwKh1o (URL 1.3)

His video simply shows him sitting in front of his computer, lip-syncing to O-Zone's song "Dragostea din tei." Across various video-sharing sites, the video has been viewed more than 700 million times and ranks as the second most viewed online video ever, beaten out only by Star Wars Kid.

www.youtube.com/watch?v=HPPj6vilBmU (URL 1.4)

You don't need money or corporate backing to go viral. You simply need a very good, or very bad, video and the know-how to get viewers to your video—both of which you will learn how to do in this book. Those viewers you bring in might subscribe and pass your video along to their friends through emails or instant messages. Or at least, that's the goal you're shooting for, and the Most Viewed page can help bring a lot of eyeballs to your videos.

YouTube has many areas to explore, but you need to register an account to do so. I'll walk you through registering your own account (which YouTube calls a *channel*) in Chapter 4, and then I'll survey more of YouTube's pages and functions. In the next section, Michael will get you started with a little more about the "level playing field" of YouTube and then teach you the basics of your equipment and filmmaking know-how.

The Joys (and Problems) of the "Level Playing Field" That Is YouTube

Michael W. Dean

This morning Alan sent me a YouTube video

www.youtube.com/watch?v=2US_SJZVsZk (URL 1.5)

of his pal nalts giving a spirited and funny rant on why mainstream media sucks for showing only the big players (the top 5 percent in terms of the number of views) whenever they parody anything about the cultural force that is YouTube. I agree with what he's saying, but it's not surprising that mainstream media works that way. Television (*ThemTube*) producers and magazine writers are busy people who get hundreds of emails a day saying, essentially, "Check this out! It's gonna change the world!" (Everyone says that, and everyone is usually wrong.)

The main job of mainstream media, other than writing and reporting, is sorting—sorting the few gems out of the crap pile that is their inbox. It's a tough job, so they tend to rely on others and go only for something that someone else has already noticed. They do not create a buzz; they report an existing buzz. So, they tend to get into a feeding-frenzy mode. Nothing is worthwhile until someone else tells them that people are looking at it. This is also true of traditional media gatekeepers, such as A&R agents at record labels, talent agents in Hollywood, modeling agents, DVD distributors, and so on, ad nauseam. (*A&R* means "artist and repertoire." A&R agents are basically talent scouts.)

In 1968, maverick pop artist Andy Warhol predicted, "In the future, everyone will be world famous for 15 minutes."

http://en.wikipedia.org/wiki/15_minutes_of_fame (URL 1.6)

Mainstream media hacks have dragged this "15 minutes of fame" statement up again and again with regard to the Internet in general and, more recently, with regard to YouTube. And it has a good point, even if it's a no-brainer. What it fails to report is that Andy Warhol got sick of being asked about the "15 minutes" and started telling people, "In 15 minutes, everybody will be famous" and "In the future, 15 people will be famous." These statements are even truer on YouTube, and the 15 people change almost every 15 minutes. (Another one that I like is "On the Internet, everyone will be famous to 15 people.")

With the exception of world-changing news events (like the hanging of Saddam Hussein, captured on a grainy cell phone and uploaded to YouTube where it was viewed millions of times), most of the stuff that gets famous and builds a following for people have a few common denominators. Most viral videos have decent storytelling, decent writing, good lighting, good sound quality, and no technical issues (like bad encoding). There are exceptions to this, like fluke videos of a tin can in a microwave or some dude wrecking his genitals while falling off his skateboard onto a railing. But while these may get hundreds of thousands of hits in a short period of time, the person uploading will have trouble re-creating that viral magic. If you want to make a go at having some longevity on YouTube, you will need to understand the basics of filmmaking, and that means understanding the basics of storytelling, writing, lighting, cameras, editing, sound, and encoding.

Don't worry, it's not rocket science. It doesn't involve math (well, nothing past eighth-grade math, and not even much of that). You can learn these basics in a day, or in a week if you're busy with a lot of other stuff. You can master them in a month. But if you don't learn these basics, there's very little chance you'll ever become a major player on YouTube.

2

Storytelling and Directing

Michael W. Dean

I know a lot about filmmaking on an indie, short-form, YouTube level and even a lot about more complicated, long-form, pro filmmaking that approaches a Hollywood level. You don't need to know all that I know to make great YouTube videos, because all of it would fill many books. (In fact, I've written, edited, or contributed to several books.)

www.cubbymovie.com (URL 2.1)

I'm going to take the more basic and YouTube-specific of that info and distill it down into the two chapters of information you'll really need to be a contender on YouTube. A lot will be covered quickly, but not in a way that misses anything you need to know to make videos that look and sound better than most of the fare on YouTube.

The basics of writing, lighting, sound, camera placement, and editing are the same, whether you're shooting with a webcam in your bedroom, making a $100,000,000 Hollywood blockbuster with a huge crew and movie stars, or shooting scrappy little documentaries that change the world with good mini-DV cameras, with a small volunteer crew mostly made up of film school students, and with a professional editor, like I did with my two feature films.

The Importance of Storytelling

Most people when they get a video camera just start shooting. I know I did. But you really should do some exploration into storytelling so you have something worthwhile to shoot.

The Basics

I cannot overstress the importance of storytelling. Storytelling at its most basic is just one person talking to another person. It predates spoken language, probably originating with the Neanderthals 500,000 years ago, grunting and emoting with their hands around a fire in a cave to describe the day's hunt.

Today, storytelling at its most basic is two people sitting on a couch talking—see Figure 2-1—not much different from the cave stuff when you really think about it.

Figure 2-1. Storytelling at its most basic, but with some recording gear added. Photo of Michael W. Dean and Mike Kelley podcasting, taken by Debra Jean Dean.

On YouTube, as with any *motion art*, the "story" is the video. Your goal on YouTube is to have you sitting on the "virtual couch" of the Internet telling your story, whether it's a story from your life or one from your imagination. The story is what grabs people.

Sure, if you're young and cute and female, people will probably be more likely to pay attention to you for a minute on YouTube. But that's not the story. And it's not really positive attention, and it isn't something that lasts. And it's not going to make you a lot of money. Storytelling is what grabs attention, keeps attention, and makes money (again, if that is your goal). Alan is making money on YouTube, but he didn't set out with profit as his goal. He set out to make good art and share it with lots of people. I don't make money on YouTube; I only want to spread art and am satisfied with that. We're speaking to both camps here. We're making the assumption that you want to make good videos and have people pay attention to them. Storytelling is the most important thing. People remember a lot of things about any piece of *motion art* (the term I'm going to use for anything from webcam YouTube videos to Hollywood blockbusters). People remember the stars; they remember the great lines of dialogue, the look and feel, the music, and more. But the main thing they remember is the *story*. Think about movies you love. You love them because the story spoke to something inside you in some important way. Think about YouTube videos you love. They have a great story, even if the film is only 30 seconds long. The story is what *happens* in a piece of motion art. It's the thing that speaks to some common thread of the human condition.

Note The best way to learn a new skill is to read up, ask questions, do real-life experiments, take notes, review your notes, and then try some more experiments. If this sounds like school to you, it's not. School tells you what to learn and when to learn it. This is combining action with thought and experience and doing it at your own pace. It's learning what you need to learn and ignoring the rest. School would never put up with this. ***This is the most important thing I'll contribute in this book, and this piece of advice is worth the price of the book by itself.*** And it comes from a guy who got in trouble for skipping school to go to the library to actually learn things. I'm now in demand in a bunch of cool worlds, and most of the kids I went to school with have nowhere jobs. A guy who used to beat me up writes for the local paper, which is cool, but the last thing I read by him was a story about how evil rock music is and why Nirvana should be banned and all their records burned. Steve Albini, who engineered the Nirvana record In Utero, also uses this "self-teaching with study, experiments, and notes" idea. Check out my video interview with him where he talks about it:

www.youtube.com/watch?v=Mw62MYwe5pQ (URL 2.2)

I've often said that a well-told story interests me even if I have no interest in the subject matter, because it speaks to something common in all people. Of all the documentaries I've ever seen, one that really stood out, *Spellbound*, was one about spelling bees, even though I have no interest in spelling bees. One of my favorite popcorn movies (big Hollywood movies that don't change your life but are fun to kill two hours and eat some popcorn) is *Happy Gilmore*, even though I hate golf.

Storytelling is different from writing, though they are related. You can tell a story without even using any words, and words are what writing produces. Some of the most compelling pieces of motion art (especially commercials) have little or no dialogue. But something always happens in motion art. Your first job as a video creator is to make things happen on the screen. This can be done by writing or otherwise visualizing a plan and then enacting that plan and filming and editing the result, or it can happen by filming things that happen without your intervention and presenting them in a compelling way with your editing. Both methods are types of storytelling.

Conflict Is the Essence of Drama

The first thing they teach you in "real" film school is that "Conflict is the essence of drama." It's true. Without conflict, movies and TV would fall out of the screen and drip lifelessly onto the floor, as this photo of my cat, Charlie (Figure 2-2), looks about to do. (The white blur is actually a second cat leaping by, but Charlie didn't catch it. She's too catatonic.)

Figure 2-2. Stories without conflict are boring. (Model: Charlie Squitten Jr.)

All movies and all TV shows incorporate story, with characters clashing with one another (see Figure 2-3), having various types of crises of conscience, and then resolving the conflict, tying it all up in a nice neat package near the end of the show. These types of crises of conscience might be called *confrontation*—in act one, the characters are introduced, the problem is defined, the story set up; in act two, it evolves, it gets worse, the conflict grows; and finally in act three, it gets resolved somehow. The same is true of most books, plays, and even songs, especially hip-hop and country music, the two most story-driven music genres in existence. The fans of the two genres may not get along, but they have more in common than any other genres (except country music probably has more guns and drugs and booze and sex).

Figure 2-3. Conflict is the essence of drama. (Models: Fuzzbucket "Fuzzy" McFluffernutter and Charlie Squitten Jr.)

This method of conflict/drama/resolution is called the *three-act format*. A common way this is expressed in Hollywood films is what is known as the *hero's journey*. Almost all Hollywood movies follow this journey—to the point that it becomes cliché. To the point that, if it's not there, the audience feels cheated, even if they can't describe what's missing. To the point that, if certain types of events don't happen at almost exactly 10 minutes, 18 minutes, 29 minutes, 46 minutes, 101 minutes..., people feel uneasy when they leave the theater. (My wife and I once had a very long discussion about whether this is because people are used to seeing it or because people need it because there's something inherent in the human experience that makes people want to organize stories into this format. The conclusion we've both come to is "It's a bit of both.")

Most hero's journeys are in three-act format, and in movies, most three-act formats contain the hero's journey, but they do exist independently of each other.

When a hero's journey is not in three-act format, it's usually a cyclical story, with parallel stories that occasionally intersect and the beginning of the film being part of the same scene that ends the film. A good example of this is *Pulp Fiction*, which still contains a hero's journey, where there is a battle in the innermost chamber and enemies become friends.

The evil baby, Stewie, on *Family Guy* (one of the only TV shows I'll watch…most TV sucks) summed up three-act format perfectly, while sarcastically deriding Brian the dog for not working on his novel:

> "How you, uh, how you comin' on that novel you're working on? Huh? Gotta a big, uh, big stack of papers there? Gotta, gotta nice little story you're working on there? Your big novel you've been working on for three years? Huh? Gotta, gotta compelling protagonist? Yeah? Got a' obstacle for him to overcome? Huh? Got a story brewing there? Working on, working on that for quite some time? Huh? Yea, talking about that three years ago. Been working on that the whole time? Nice little narrative? Beginning, middle, and end? Some friends become enemies; some enemies become friends? At the end, your main character is richer from the experience? Yeah? Yeah?"

So, yeah, the three-act format is a cliché but one worth understanding. And by the way, once I explain it to you, you'll never look at Hollywood movies quite the same way. You'll feel like you're being lied to. Because you are. Life is not that neat, things are not always completely resolved, and every situation does not have a lesson or a silver lining. Many do, though. I have a pretty positive outlook on life, but I hate sugar coating. In any case, here you go…

The Hero's Journey

A hero's journey story has three acts. Three distinct feelings. The first act is 30 minutes, the second act is 60 minutes, and the final act is 30 minutes. If the movie is longer or shorter than two hours, adjust accordingly. But regardless, the second act is about twice the length of each of the other two. This is shown in Figure 2-4.

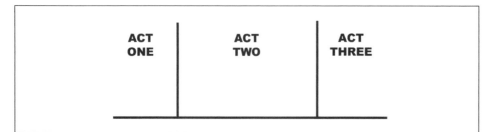

Figure 2-4. Approximate division of time per act in three-act format.

The first act takes place in the "normal world," the world the protagonist (hero) normally lives in. During this first act, we meet the hero, find out his (or her) problem, and see his *call to action*. This is him being called to act heroically, in a way that would lead to the rest of the story. *The hero always refuses the call the first time.* Then something changes, it becomes personal, and he has *no choice* but to answer the call to greatness and risk it all to become a hero. The hero is usually played by Brad Pitt, Bruce Willis, or Wesley Snipes.

The hero meets a mentor, an older person who used to be at the top of the same game the hero wants to operate in, but the mentor is now retired or crippled in some way, so all he has to pass on is knowledge and wisdom. He also usually passes on some sort of talisman—an actual physical object—to the hero. The talisman seems useless at this point but will help the hero in some way in the third act. The mentor is usually played by a handsome, older black guy with a deep voice, and it's usually Morgan Freeman.

The mentor helps the hero assemble a ragtag crew of misfits as his team to help him in his hero's journey quest. There's always one guy on the team who seems like he's going to get everyone killed. This is the guy who makes you think, "Why is *that* guy on the team? He's going to get everyone killed!" This character is always played by a short, squirrelly, funny-looking white guy, often Steve Buscemi.

Once the hero's team has completed its training, the assembled team will always walk in semi-slow motion side by side toward the camera, wearing whatever outfits they will wear in their new world. This shows that they're now a single unit, not four or five individuals. This badass walk is how you know act one has ended and act two is beginning.

If it's a movie about hookers, they're dressed in pumps and miniskirts. If it's about astronauts, they're wearing space suits. If it's Reservoir Dogs, they're wearing cheap suits and skinny ties.

The second act takes place in the "special world," the world of wonder where most of the story actually happens. This can actually be a different physical location from the normal world, or it can take place in some marginal society that's still geographically in the same town as the normal world. But it's often a different physical location. And when it is, part of the "journey" will involve the hero and his team traveling to that physical location.

In this special world, the hero will battle many foes and almost die several times (figuratively or literally). He will become stronger, physically and spiritually, with each battle, until he meets the ultimate evil (antagonist) in the third act in the *battle in the innermost chamber*. This is the climax of the film, and it usually happens in the middle of act three. The most quotable lines from the movie come from this part. The quote on the movie poster often comes from this part.

If it's a cowboy movie, the battle in the innermost chamber takes place in the corral as a shootout with the villain. If it's a courtroom drama, it is closing arguments screamed by lawyers and defendants in front of a jury while the judge bangs his gavel and demands order. If it's that dodgeball movie, the battle is the dodgeball championships in Las Vegas. During the battle in the innermost chamber, the hero will die (figuratively or literally), resurrect with wounds (this is probably a Jesus metaphor), and stand again to finally slay the beast, even though he's bleeding (figuratively or literally).

Then there's the denouement (pronounced "day nu ma"), the gentle anticlimax where the hero gets the girl, gets the gold, saves the farm, and returns to the normal world. But now he's richer for the experience, changed for the better, and bearing gifts (literal or symbolic) to help his community. (This section often leaves one or two seemingly small questions unanswered, which sets the story up for a sequel, just in case the film makes a lot of money.) Roll credits.

Do you feel cheated? You should. Hollywood relies on this formula to make safe movies that are guaranteed to turn a profit. They even wrap this formula around true events that do not occur in three acts, in biopics and documentaries. Sometimes, especially in reality TV, the conflict is even manufactured, put together behind the scenes by the producers. (See Figure 2-5.)

Figure 2-5. Sometimes there is conflict.

No cats were harmed or even mildly irritated in the making of these images. They illustrate the idea of contrived conflict, and it does look like the producer is strategizing behind the scenes to create conflict. It looks like Debra Jean is putting one cat near the other cat, but she's actually gently removing the cat who was about to fight, thereby stopping a fight.

Scripts that don't follow the hero's journey recipe to a T are routinely turned down by producers, even if the story and writing are great. Hollywood has so much potential, and it's almost always wasted. It's sad. The hero's journey does, however, have some good attributes. Study this formula, know it when you see it, and then rewrite the rules. Pick the parts you like, discard the rest, and make something brilliant. (Or do what *South Park* does and follow the formula to the letter, exaggerate all the bullet points, and make fun of it while you're doing it.)

Most good videos have a discernable beginning, middle, and end; even unscripted *vlogs* (video blogs—confessional, one-person-looking-into-the-camera-and-baring-his-soul-type videos, which are very popular on YouTube) have these three parts. The best vlogs have a feel of introducing an issue, talking about the issue, and resolving the issue, and the best vloggers do this without even thinking about it.

A Likable Main Character

Every story, no matter how short or unconventional, needs at least one likable main character. This character doesn't have to be perfect—he can actually be quite flawed—but there has to be someone for the viewer to "root for" and live vicariously through.

I figured this out recently. My first novel, *Starving in the Company of Beautiful Women*, did not have a likable main character. In fact, the main character dies on page 1, and then the book is all back story. While it's hard to cheer for a guy who's already dead on page 1, this might have worked if I'd made him a little more likable. The cult film *Liquid Sky* has no likable characters.

> *www.youtube.com/watch?v=xwOHKKJxrgA&feature=related* (URL 2.3)

> *www.youtube.com/watch?v=S9-n9gpFVpk&feature=related* (URL 2.4)

It's a very interesting movie otherwise, especially visually and sonically, and probably would have had a wider audience if there had been someone to root for. But the characters don't even care about themselves, so why should we care about them?

The need to have a likable character is even seen in nonfiction coverage of stories with characters you can only despise. In every news show that covers the brutal murder of a family in a home invasion, they always interview the crying surviving relative. That's the person you root for in that story; they're likable because you feel sad for them.

Brevity Is the Soul of Wit (or 2 Minutes to Fame)

Basically, as a filmmaker, you have about 2 minutes to get the attention of your viewers. Maybe less. Here I'll tell you why.

Hit 'Em Over the Head and Get Out

YouTube currently allows you only 10 minutes to tell each story anyway. It's pretty hard to make a fully rendered hero's journey story in 10 minutes. It can be done, but you'll be pushing it and skipping over the development of character and plot that requires 90 minutes or more to really do well. I recommend going with *two-act format*, which works really well on YouTube, especially for videos less than 5 minutes. (And most videos that go viral are less than 5 minutes, because people have really short attention spans these days.) With a short-form two-act format, you're basically hitting 'em over the head, then getting out quickly.

 Alan spends more time on YouTube than anyone I know, so I believe it when he says, "The average length of a viral video is 2 minutes."

The two-act format is basically a setup and a knockdown. A "Knock knock" and a "Who's there?" A joke and the punch line. As you see in Figure 2-6, this often works out to 90 seconds of setup and 30 seconds (or less) of punch line.

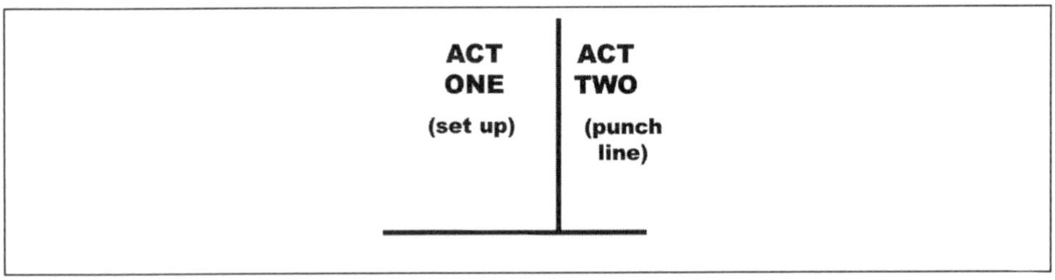

ACT ONE
(set up)

ACT TWO
(punch line)

Figure 2-6. Approximate division of time per act in two-act format.

Another way to do it is 90 seconds of setup, 15 seconds of punch line, 10 seconds of denouement, and then another quick punch line after you feel resolved and relaxed. (See Figure 2-7.)

Even if your video is not humorous, a two-act setup and knockdown works great on YouTube, on the small screen, with tight time limits.

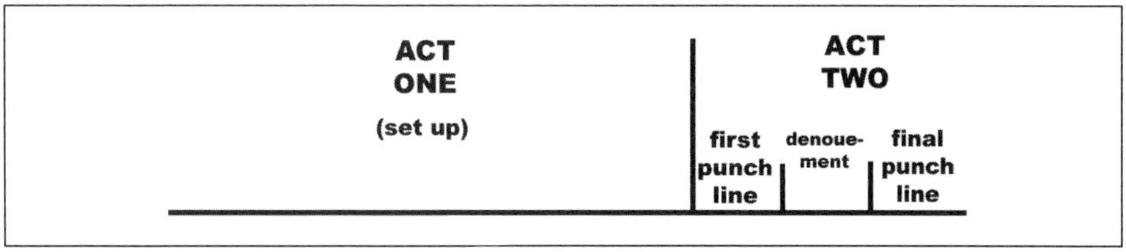

ACT ONE (set up)	ACT TWO

Figure 2-7. Another approximate division of time per act in two-act format.

Think about the audience of YouTube. It's mostly the post-MTV generation—folks who want action, change, fast pacing, and a quick bang for their (free) buck. It is not usually the same people who watch long, slow, black-and-white art films with subtitles. And even though some people like that (like me) are viewers and even content creators on YouTube, when we're on YouTube, we don't really have that mind-set. For many people, the Web in general, Web 2.0 (social networking like YouTube, MySpace, and Facebook) in specific, and YouTube even more specifically are used as brief time killers between other tasks. People know that they can go to YouTube, click a bit, and be made to laugh, cry, and laugh again in less than 5 minutes. From watching three complete videos. That's more or less the mind-set you're playing to on YouTube.

Think about TV commercials. YouTube videos, at their best, are written with a pacing similar to that of TV commercials. TV commercials have 30 to 60 seconds to do it all. In this amount of time, a commercial has to grab your attention, introduce a product or service, make you understand the product or service, and make you care enough to part with your hard-earned money. They have to mention the name of the product or service three more times so you'll remember it and then tell you where to go to get it. That's a lot to accomplish in half a minute. I hate TV more and more, but more and more, the thing I still find watchable on TV is the commercials. That's because the most creative minds graduating from film and art schools usually get bought up by ad agencies, because there is so *damn much money to be made* in advertising.

Note I recommend that if you really want to understand advertising, and the world, you should watch the 1988 John Carpenter film *They Live*:

http://en.wikipedia.org/wiki/They_Live (URL 2.5)

This 1988 low-budget dark comedy has a sci-fi motif, but it has more truth than most documentaries. *They Live* shows the world as it really is behind the matrix and has a lot of great stuff to say about advertising. You can get a copy of *They Live* used on Amazon for seven bucks plus shipping:

www.amazon.com/They-Live-Roddy-Piper/dp/B0000AOX0F (URL 2.6)

In general, I hate the idea of ads on TV, in magazines, on social networking sites (particularly MySpace, which is starting to look like TV—or more like four TVs going at once), or anywhere. I don't buy things based on ads, and I consider ads obtrusive. When I want to buy something, I shop bargains, ask friends, and check user reviews. Check out these TV ads for milk, featuring an imaginary milk-fueled rocker named White Gold. These are some of the coolest bits of motion art ever created (and they're ads for milk, a product I enjoy and actually consume a lot of). This is the story of how White Gold came to be: *www.youtube.com/watch?v=fVXHRC0DG70* (URL 2.7)

This is the best one, his song "One-Gallon Axe." It even has a llama and a Viking helmet! *www.youtube.com/watch?v=nj1-ZZRajDU* (URL 2.8) The guy is the singer in a real band in L.A. called The Ringers. They're great:

www.youtube.com/watch?v=GF_k-YYCpno (URL 2.9)

The style and pacing behind advertising is a great way to get things done, get people's attention, and make them remember something. Study commercials to learn how they work and then learn to subvert the tricks of advertising to serve your own agenda.

Slower Brevity and Wit

This "rule" of "Get it done fast, make your point, entertain, get out fast" is not hard and fast. Many vlogs work well with a slower, more laid-back pace, and they still captivate your attention. This vlog by Alan called *50 Things* (see Figure 2-8) is slower yet still very compelling: *www.youtube.com/watch?v=LjfL5WL4ZBQ* (URL 2.10)

Figure 2-8. Alan vlogging from the heart.

This video manages to work well in a relatively short time frame, because the subjects covered are interesting, seamlessly going between intimate and funny, and Alan has a good, very natural, delivery. He's confident without being cocky. The mood is interesting, the feeling is from the heart, and it is both uplifting and personal.

A cocky mood for delivery can work also in vlogs and other types of "personal" videos. I don't personally like this style as much, but a lot of people seem to dig it. Here's a good example from user DoggBisket:

www.youtube.com/watch?v=q_wCbwIwfQg (URL 2.11)

This is also a good example of an effective collab video, involving several people in addition to DoggBisket, including Alan. It also credits everyone well in the credit roll at the end. This is going above and beyond; most people just credit in the More Info link on the top right of the page. Crediting in the video itself is a nice touch for the people who work so hard for free to help out.

Keep in mind that the basic mood, and the audience, of YouTube is highly influenced by television commercials, rock videos (which are basically commercials for music), and even talk radio, as you can see in the video by DoggBisket. If something takes 8 minutes to say but could be said in 4 minutes, 4 minutes is probably better. Johnny Cash said that if a line of lyrics doesn't advance the story in a country song, cut that line. An old punk-rock nugget of wisdom from back in the day was "Why have a guitar solo when you can just make the song shorter?" and, with the occasional exception, I tend to agree.

You don't need to chop your videos down to 60 seconds, but if you keep in mind the short attention span of the average YouTube viewer while writing and editing and go for punch rather than long, sweeping explorations, you'll have a better chance of achieving the coveted "going viral" goal.

I recommend you start with some 5- or 8-minute projects and then work on shaving future ones down to 2 minutes for your better ideas after you get really good at writing, shooting, directing, editing, and promoting. Remember: 2 minutes to fame.

The Importance of Writing

You will need to write some for your story before you make a video. This can be as elaborate as writing a word-for-word script, writing a short outline, or even just writing it in your head, but the better videos all have some planning behind them.

Treatments and Scripts

Writing is related to, and part of, storytelling (and vice versa), but it's so different that I decided to give it its own heading, one that's equal to storytelling.

> **Note** See what I just did there? I pointed out that I am writing this. That's called *breaking the fourth wall* in filmmaking; it's when an actor stops acting and addresses the camera directly, in a way that "admits" there's a camera. It can be an effective tool, when used sparingly, and for a planned mood, in drama and comedy, and is pretty much the basis of all vlogging by definition. Breaking the fourth wall is your first lesson on writing for motion art. Enjoy.

Storytelling is the broad strokes. When a Hollywood writer pitches a story to a studio, they often have only the brief outline of a story. There's even something called an *elevator pitch*, which is supposed to be the shortest amount of time it takes to convey a complete story, like if you found yourself sharing an elevator with a producer for a ride of only a few floors. This is also called *25 words or less*. The plot of most Hollywood films can be completely expressed in 25 words or less. For YouTube, think 15 words or less. This is true of most Hollywood films. Complicated ones that break the hero's journey mold (and these are usually my favorite films) like *Fight Club* and *Memento*, still have some sort of three-act structure, albeit convoluted.

Once a Hollywood movie is sold, based on the pitch, and the writer is hired, they're often asked to produce a *treatment*. This is a longer pitch (5 to 20 typed pages), but not quite as long as a full movie script (about 105 to 120 pages, which works out to about a minute per page in the final product). The treatment explains all the main characters, their ages, their names, a little background, their motivations, their dreams, their desires, their strengths, and their flaws. Then it has rough thumbnail breakdowns of the major plot points in the story. The full script is written using the treatment as a guide, after the treatment is approved (and usually meddled with) by the studio. Since YouTube gives you a maximum time length of 10 minutes, your treatments will not be nearly this long, if you even write treatments at all.

> **Note** Director accounts registered before January 2007 have no time limit, only the 100 MB basic file size limit that everyone has, or up to 1 GB if you use the Multi-Video Upload feature. Currently, all new accounts have a 10-minute limit to deter TV show and movie uploads.

Many YouTube viral videos do not have any kind of scripted, or written, aspects— they're things like major news events, people hurting themselves in sports accidents, or a kitten doing something incredibly cute. But while these videos may get a million views, they don't get many subscribers, because they're so *one-off*. Here's a 50-minute video on the history of YouTube that explains how the site grew: *www.youtube.com/watch?v=nssfmTo7SZg* (URL 2.12) Assuming you bought this book to build a following, you probably want to go for some form of scripted motion art.

How to Write an Outline

Alan does a lot of *collab* videos. (Collabs are collaborations—videos done with other people, usually over the Internet. More on this in Chapter 8.) One person Alan has done a lot of collabing with is Lisa Nova (username LisaNova). Lisa's user page is here:

http://youtube.com/user/LisaNova?ob=1 (URL 2.13)

Lisa is one of the bigger rock stars of YouTube. Not only is she in the top 10 all-time most subscribed directors on YouTube, her work on YouTube led her to being a regular on MADtv for a season. She's now getting so much work outside YouTube that she's actually hired Alan. This is pretty sweet, and it's one of the side benefits of doing amazing stuff on YouTube and making it great, even when it's only a little fun parody video. That can grow into more. A lot more. One of the collabs Alan and Lisa did early on was this parody of an iPhone ad (see Figure 2-9): *www.youtube.com/watch?v=ZpUItDR5im4* (URL 2.14)

Figure 2-9. Screenshot of LisaNova from the iPwn video.

Alan wrote the concept and the *copy* and emailed it to Lisa. (Copy here means the words that the actor reads. *Ad copy* is the words actors in a commercial read.) Danny, Lisa's producer and partner in art, shot a few takes of Lisa delivering the copy and sent the footage to Alan who had my wife, Debra Jean Dean, *www.debrajeandean.com/* (URL 2.15) do the little "in time for Christmas" voice-over at the end. I engineered it and emailed the audio file to Alan. A script can be as easy as a single email.

Alan's script consisted of one email, which covered everything. In that one communication, Alan asked Lisa and Danny to participate, outlined the concept, showed the copy, and even included a minimal bit of direction and a salutation and mention of a related project ("the Stickam idea"—doing "making-of" videos of this shoot through the live video site Stickam.com).When you're working with talented people, that can be enough. Here's what Alan sent Lisa, care of Danny.

Danny,

A collab with Lisa would be awesome; she actually mentioned it in her comments on our videos yesterday. I have one idea, but it would basically be me writing for Lisa, I wouldn't appear in the video.

It would be a mock commercial, advertising Apple's latest advancement, an elite wireless phone/gaming system called the iPwn. (Pronounced "eye pown," I'm sure you're familiar enough with gamer/chat speak.)

Anyway, it would be her as a gamer girl, standing in front of a plain background (black or white) talking about her new iPwn... "Not only can I email my friends, get GPS directions, and listen to the latest Soulja Boy remix, but I can also frag n00bs effortlessly with its responsive touch pad display."

Etc. I have a professional voiceover actress to do the end tag line, and I'd create all the graphics, etc.

What do you think?

Here's a link to the original iPhone commercial:

(You've seen the iPhone commercial on TV so you know what I'm going for.)

I'd like to keep the framing, and mood, lighting, etc., similar.

Continued

Here's the script:

"One night, I'm a level 24 Paladin and my alchemy experience is all that matters. The next, I'm known for my double kills and stealth mobility. I need a wireless gaming device as diverse and adaptable as I am.

"That's why I'll only use the new iPwn.

"Not only can I email my allies, get GPS directions and listen to the latest Soulja Boy remix, but I can also frag n00bs effortlessly with its responsive touch pad display.

"I'm a girl gamer. And if it weren't for my new iPwn, I'd be lying dead in some enchanted ditch."

And if I don't answer, please leave a voice mail, I'll usually call right back, I don't answer numbers I don't know, too many international computer weirdoes making prank Skype calls.

And thanks again, guys, the Stickam idea sounds insanely cool.

Alan

Alan did "remote directing" via live video site Stickam.com and used a webcam to actually give Lisa direction as she shot her part, 2,000 miles away. It worked out well. If you want to see behind-the-scenes, it's here:

www.youtube.com/watch?v=d76vEPhDX7E. (URL 2.16)

Choosing and Directing Actors

Actors are the talking puppets who bring your stories to life. OK, they're more than talking puppets; they're humans, with real emotions and feelings. Either these emotions and feelings can get in the way of putting your story on the screen or you can work with these emotions and feelings…and desires. That's what a director does— "directs" the actors to channel their human experience onto the screen.

Choosing Actors

Most YouTube videos star amateurs as the actors. In fact, most YouTube videos star you and your friends. (That's why it's called YouTube, not ThemTube. *ThemTube* is our nickname for television.)

Figure 2-10. YouTube. Yay!

Figure 2-11. ThemTube. BOO!

For vlogs, and for funny, cute stuff, having you and your friends is fine (Figure 2-10) and it sure beats *ThemTube* (Figure 2-11). But what if you have a more sweeping vision in mind? What if you want to make something "as good as Hollywood," but better? You will need some people who can act.

For all the things that Hollywood does wrong, it does a lot right. What they do wrong is this: They tell the same hackneyed hero's journey, happy ending story over and over. The situations may change, but it's pretty much mostly all the same movie. This is called *pandering to the lowest common denominator*, and if you're smart, you should probably be insulted by it.

What Hollywood does right is this: Everything is technically perfect. The sound is amazing, the cinematography is stunning, the editing is great, and the actors are really, really good. Sure, they may feature actors who are cute women more than actors with a lot of experience and talent, but even the cute women hired mostly for their looks can act.

When I say "They can act," there's something you need to understand, and that is "what acting is." Acting is *lying*, in a creative way. Acting is convincing the camera, and the audience on the other end of the camera, that you are someone you're not. It sounds a little sinister when put in those terms, but it's really not. It's an age-old art that predates YouTube, predates movies, and even predates the Greek plays of 400 B.C. It probably goes back to the first-ever caveman sitting around the campfire doing impressions of the second-ever caveman for the amusement of the third-ever caveman. We'll never know, because there was no way to record anything then (that's why that era is called *prehistoric*—before history), but I'd say that's a safe bet.

Not everyone can act. Some hopefuls spend years and thousands of dollars on lessons, seminars, and books, and they still suck. Other folks are a natural and shine the first time a camera is pointed at them. The best actors are usually both. They've been doing skits and making people smile and believe since they were little kids, and they also have some formal training.

An important skill as a director, even on the tiny screen of YouTube, is to be able to recognize great acting. Another large part of what a director needs is people skills, and that includes the skill of being able to say "no" to someone, even if they're your friend.

I don't mean to sound like a hard ass. And your first few videos will probably star everyone around you—yourself, your family, your girlfriend or boyfriend, your neighbors, and even your cat. But as you get a lot better at writing, directing, filming, and editing, you're going to want to find better actors. You're telling stories, and you want people to believe the stories and be lost in the stories, so you don't want bad acting to get in the way, any more than you want bad audio to get in the way.

A great way to make yourself a better director is to know good acting when you see it. And a great place to start is to know *bad* acting when you see it. Then, motivate your actors to do the opposite.

Here's a clip titled *Bad acting 101* on YouTube, and it certainly is:

www.youtube.com/watch?v=qBcpW-7awMA (URL 2.17)

The thing that I would say defines bad acting is "not being believable." And the worst problem most people have with not being believable is "overacting." That is, acting too *big*. Being too enthusiastic. Being too loud. Moving your arms and head too much. Bad actors often act like they're on a stage in a play and need to exaggerate everything for people 100 feet away in the back of the room. Good actors realize they're acting for a camera a few feet or even a few inches away, and they bring it *down*. Smaller and realer. Great actors are amazing to watch. They reach through the camera and touch your soul. They *become* the person they're playing.

 It's scary when someone is mad and they get louder and louder and yell a lot. But it's even scarier when the madder they get, the quieter they get. It's reptilian. Suggest this when directing actors, or try it yourself when directing yourself as an actor.

Here's some good acting (as well as great cinematography and great directing). It's a film by my friend Noah Harald who taught me how to make films, back in 1998.

www.vuze.com/details/VGM52SKAUPNE4YU56XNHLWC6E4A2PFAO.html (URL 2.18)

 Noah actually makes a living making films. Here are his resumés:

www.noahharald.com (URL 2.19)

His stuff is amazing. When he was on YouTube, he didn't get many views, which illustrates something I find lacking in YouTube…. Diet Coke mixed with Mentos erupting gets upwards of a million views:

www.youtube.com/watch?v=hKoB0MHVBvM (URL 2.20)

Noah was nominated for an MTV award, he was on MTV at the awards show, and MTV showed his short on there, among a lot of other stuff. He is a graduate of Los Angeles Film School and has a paid webisode deal with Filmaka.com:

www.filmaka.com/filmaka_series.php (URL 2.21)

Here's some good acting, on a video that's gotten a lot of hits on YouTube:

http://youtube.com/watch?v=CeAtbrYVqqs (URL 2.22)

Sometimes, especially on the short-attention-span medium of YouTube, bad acting can work, if the people are actually good actors and are "playing" bad actors. Here are examples of bad (in my opinion) acting, where they're not ashamed of it, where the bad acting "works":

Daxflame: *http://youtube.com/watch?v=yUwVqL1xtVs* (URL 2.23)

thebdonski: *http://youtube.com/watch?v=MrZPxMoB92E* (URL 2.24)

(thebdonski is LisaNova's brother, who is actually a fabulous actor who's appeared in a few feature films but plays it down here…or plays it up, depending on how you look at it.)

Directing Actors

Directing can include everything that leads up to what you see on the screen. In Hollywood, some directors mainly concern themselves with directing the actors. Others are more into directing the cameraman and just let the great actors do most of it themselves. Other directors are total control freaks who reign like kings over the acting, camera work, set design, music, editing, and so on. You can choose how much of what you want to do (and in many cases, you'll be doing some of this yourself, like probably the camera work, so the directing will be all within your head).

Note I **strongly** recommend you subscribe to this YouTube channel:

www.youtube.com/user/esotericsean (URL 2.25)

and watch all the episodes. It's a filmmaking instructional series called *Take0*, which refers to everything you do before the first take. They teach all about planning, although they also cover what to do **while** shooting, and after, to make a great short film. They cover writing, planning, directing, acting, camera work, editing, and more. Two young smart guys make it, and they add new episodes regularly. They have more than a thousand subscribers.

I'm honored to know them and honored that they cite my *$30 Film School* book as an influence.

Motivation

Your first job as a director is to motivate the actors. Give them suggestions to bring the script to life. Or if it's a looser, more improvised production, bring the outline to life. The best way to do this is to gently but firmly tell them what you want. The archetype of the screaming director is not one to emulate and, especially if you're dealing with unpaid amateurs, will more likely result in people quitting with tears streaming down their faces than it is to result in a good performance. (See Figure 2-12.)

Figure 2-12. Michael directing and having a bad hair day.

Try to coach your actors into giving you what you want. You have a vision; they are there to help bring that vision to life. Different approaches work for different actors. Some respond to sweet talk, some to a more emotionless pragmatic approach, some to being told to "picture themselves as the character." A lot of directing is simply people skills, getting great work from people and sussing out the best way to do this with different people. Practice this, and keep track of what works and what doesn't. You'll get better with time at figuring out the best approach, to the point where it becomes intuitive the first time you meet someone and talk to them for a few minutes.

There's effective video art that comes from a group directing mentality but I've found it's not as good, and results in some degree of chaos, if one person isn't in charge. If you have a clear vision and your actors are chiming in suggestions that aren't helpful, ask them nicely to follow the script and your direction. If they aren't happy with this, find different actors. If they're decent actors but trying to run the show, and they're also YouTubers or filmmakers themselves, you can say, "If you do this my way in my film, I'll be an actor and let you direct me without commenting in your film." Being directed by someone else, even a bad director, can make you a better director.

Directing the Camera

I saved the most fun part for last.

The other main part of what directors do, other than directing the actors, is directing the camera operator. The camera operator might be you in your YouTube videos, or it might be a friend. Regardless, you should understand the shots that make up most filmmaking.

Shots

The basic shots in Hollywood filmmaking are the establishing shot, the medium close-up, the close-up, and the over-the-shoulder shot. These shots are shown in Figures 2-13 through 2-17.

An *establishing shot* is a wide shot that shows the setting where the scene takes place. A lot of beginning filmmakers skip this, and that can be a mistake. If you start with a close-up of two people talking, the viewer has no frame of reference for where the scene is taking place. This is not always needed; sometimes it's all about the people. But in any story-driven piece, consider whether the location is important. If it is, you might want to consider starting the scene with an establishing shot.

Figure 2-13. Michael and Debra Jean in an establishing shot. This photo and the next four photos taken by London May. ©

A *medium close-up shot* shows a bit of the background of the setting but comes in closer on the people talking. This is often where you "meet" the characters for the first time.

Figure 2-14. Michael and Debra Jean in a medium close-up shot.

A *close-up shot* shows more of the mood of the face of one character. It should be used sparingly because it can be a little intimidating. It is often good for showing a reaction, especially of shock or surprise.

Figure 2-15. Michael in a close-up shot.

An *over-the-shoulder shot* is often used to show the face of the person speaking, while putting it in the context of the other person listening. It can be overused and grows tiresome quickly, especially if you switch angles every time each person speaks.

Figure 2-16. Michael and Debra Jean in an over-the-shoulder shot, shown from over her shoulder.

Figure 2-17. Michael and Debra Jean in an over-the-shoulder shot, shown from over his shoulder.

Here's a great tutorial on using different shots (see "more info" on the right to see which shots show up at each time point): *www.youtube.com/watch?v=CpPWdgGa28k* (URL 2.26). You can also use different camera angles to accomplish different feels. Here's a good tutorial (from the Take0 folks) on camera angles: *www.youtube.com/watch?v=VLfI4DQIAaY* (URL 2.27)

Tripod or No Tripod?

You can buy an inexpensive tripod for your camera. A tripod stabilizes the camera. (You can also just set your camera on a stack of books if you don't have a tripod.) Shots from a stabilized camera can look more professional, or they can look stale. Handheld shots (where you're holding the camera rather than stabilizing it with a tripod or other object) can look exciting and add energy, or they can feel jittery and amateurish. Figure 2-18 and Figure 2-19 illustrate tripod and handheld camera use.

Figure 2-18. Mini-DV camera on a tripod.

Figure 2-19. Handheld Mini-DV camera.

Here's a video that was taken using handheld camera work that looks good: *www.youtube.com/watch?v=7_kWmZG1zdE* (URL 2.28). It's Alan walking around his apartment, vlogging while holding the camera.

Experiment with both stabilized and nonstabilized camera work, know what each can do, and add it to your list of tools to get the exact right look for every moment you commit to video. As the director, you are calling the shots, literally, whether you're directing someone else with the camera or doing it yourself. It's all part of getting your vision onto the tiny screen and potentially out in front of millions of people.

So, there's your crash course in storytelling and directing. Got that? Good. Time to move on to actually committing your stories to the camera and then getting them up on the 'Tube.

3

99-Cent Film School: Shooting, Editing, and Rendering

Michael W. Dean

Choosing Your Weapons

To have a chance at being viral, your YouTube videos must have some quality to their look, feel, editing, and audio. They don't need to have stellar technical qualities, but they need to have those considerations not distract from the story and the acting. People don't need to say, "What stunning cinematography!" when they see your video, but if they say, "It's blurry, and the sound is horrible," it will probably never go viral.

Here we'll talk about the first thing to consider in this: choosing your weapons, that is, picking and buying a camera, microphone, and computer. We'll cover the different formats for different budgets—mini-DV, cheap webcam, video with still camera, and video on cell phone, as well as how to get good-looking shots out of any of them and a bit on making video from stills.

Mini-DV

Mini-DV filmmaking turned the world of motion art on its ear when it was introduced in the late 90s. Mini-DV gave filmmakers something that looked almost as good as expensive film, made it cheap to shoot hours and hours, and made it editable on a home computer.

The first mini-DV cameras cost a lot, as much as $4,000, and you can still pay that much (or more) for professional-grade three-chip mini-DV cameras. But the price of consumer cameras (one-chip) has come way down.

One recommendation I have is the SONY DCR-HC48.

http://tinyurl.com/52ytnk (URL 3.1)

This unit costs about $400 new and about $300 used (give or take, depending on where you look and what the market is like that day).

Alan uses this model; it's good enough for him, and his stuff goes viral.

> **Note** This camera does not come with a FireWire cable. It does, however, come with a docking station that will allow USB or FireWire to connect with it, and it comes with a USB 2 cable. USB 2 transfers data almost as fast as FireWire.

You can get better audio if you plug an external mic into a camera, but you can't do that with the DCR-HC48 because it does not have a mic input.

The Sony DCR-HC96 mini-DV camera (see Figure 3-1) is no longer manufactured, but it has a mic input, is a great camera, and is widely available used from eBay or Amazon.

http://tinyurl.com/4bdcmz (URL 3.2)

Figure 3-1. Sony DCR-HC96 Mini-DV camera.

I generally don't recommend buying cameras used, if you can afford it. When you do that, you're probably inheriting someone else's problems and a lot of wear and tear. The problem, however, is that this good camera is no longer made. It seems that

Sony, and other camera manufacturers, have quit making one-chip cameras with mic inputs. I think the idea is that they want to sell only low-end (less than $300) and high-end (more than $3,000) cameras, with little or no middle ground, because offering the middle ground was keeping people from buying the expensive pro models. So if you want to buy a used camera, I recommend buying it from a reputable dealer, one that offers a return policy of at least 30 days on used gear. If big problems show up in used gear, they are, in my experience, likely to show up in the first month of use. If they make it past that, they're probably good for a few years more.

For external mics, I recommend the Audio-Technica ATR35S as a good cheap ($23) lavaliere (clip-on) mic.

http://tinyurl.com/4fzk7g (URL 3.3)

The Audio-Technica has decent sound, and the price is certainly right. Lots of online stores sell them. Search for *Audio-Technica ATR35S* on Amazon or eBay, and get a new one. They rock. You can plug them directly into a mini-DV camera, clip it unobtrusively onto the shirt collar of anyone you're shooting (including yourself), and get much better sound than you would get with the onboard mic.

The Basic Features of Most Consumer Cameras

Most consumer cameras include at least the ability to zoom and some kind of shot stabilization. Many also include night-vision capacities.

Zooming

Zooming is the ability to zoom in, that is, twist a dial or push a lever to bring your shot in tighter. This is good for framing your shot. (These camera shots are related to, but slightly different from, the establishing shot, over-the-shoulder shot, and other shots discussed in Chapter 2. Those are part of the language of storytelling; the shots we're talking about in this chapter are part of the mechanics of using a camera. For instance, not all wide shots are establishing shots. A wide shot can be used to establish location, but it can also be used to end a scene. And not all zoomed-in shots are close-ups of actors. You could show a zoomed-in shot to indicate detail on an object, or purely for artistic considerations that have little to do with plot or storytelling.) Figures 3-2, through 3-5 show the finer points of zooming.

Figure 3-2. Shot zoomed out wide.

Figure 3-3. Shot zoomed to medium.

Figure 3-4. Shot zoomed in close.

Figure 3-5. Shot zoomed in very close.

Shot stabilization

Shot stabilization is built-in software and hardware that makes jittery handheld camera work appear less jittery. The amount of stabilization differs from camera to camera, so experiment with your own camera to find how it works, and consider using it when doing handheld camera work. Not all lower-end cameras have this option. The SONY DCR-HC48 does not. The SONY DCR-HC96 does. (On Sony cameras, it's called "Super SteadyShot.")

Shot stabilization can also be done in *post* (postproduction or editing) with software. Here's a video with an excellent side-by-side comparison of before and after:

www.youtube.com/watch?v=aX-UxxM_6sI (URL 3.4)

This was done in post, but the effect is pretty much the same when it's done in the camera. Here's a page with more info about postproduction, or editing:

www.doceo.com/pro_editors/pro_corrective.htm (URL 3.5)

Night vision

Night vision is just that…the ability to shoot in settings with very low light. It produces a creepy "security camera" feel that is mostly green, black, and white. Not all cameras have night vision, and the location of the control for it varies in those that do. (Night vision is called "NightShot" on Sony cameras. NightShot is Sony's trademarked term for night vision.)

Here's a video that shows night vision well:

www.youtube.com/watch?v=2hDXFx-vxr0 (URL 3.6)

Shooting Video on a Cheap Webcam

I recently purchased a Logitech QuickCam Communicate STX Webcam.

www.newegg.com/Product/Product.aspx?Item=N82E16826104066 (URL 3.7)

It's not perfect, but it's a perfect way to start shooting on YouTube if you're on a very limited budget. And it's better for vlogging than for shooting stories.

My total out-of-pocket expense was $46, with shipping. It's visually a good camera, but the included software, which you must install in order to use the camera, messed up some of the sound drivers in my computer, and it took a bit of work to get it right. I recommend that when you install it, you choose the custom install feature and opt not to use the sound drivers with the camera. Just record audio separately like I do; you get much better sound that way anyway.

This camera has a built-in microphone, shoots 640 x 480 30 frames per second (fps). (Thirty fps is the same as mini-DV, although the 640 x 480 pixel image size is lower than the 720 x 480 size of mini-DV. And the color capture and overall image quality is higher with mini-DV.) Basically, if you light yourself well, you can get somewhat decent footage from it. If you don't use an external mic, don't mount the camera on or near your computer; if you do, it will pick up the noise from your computer fan.

The Logitech QuickCam has a USB interface and works on any recent flavor of Windows, including Vista. It comes with a basic capture and editing utility, or you can record directly into another program (like Sony's Vegas Movie Studio, which will be discussed later in this chapter). Or you can capture using the included basic video software and then import your footage into your editing program.

There are cheaper webcams on NewEgg.com. Check out the user reviews before buying a webcam (or anything) on that, or another, site.

www.newegg.com/store/SubCategory.aspx?SubCategory=152&Tpk=web+cam
(URL 3.8)

Shooting Video on a Still Camera

Many still cameras (that is, normal "nonmovie" digital cameras for taking still images) have the added feature of being able to shoot some video. It doesn't look fantastic, but it can look pretty good, especially if you're shooting close up and there's not tons of motion. Here's a video called "Kitty Lickin' Good" (Figure 3-6) that I shot on my Canon PowerShot still camera. My wife, Debra Jean Dean, and I did the music for it:

www.youtube.com/watch?v=Bc1eW7D42t4 (URL 3.9)

Figure 3-6. Screenshot from "Kitty Lickin' Good."

I shot it with my still camera, which also shoots video with audio. I attached my camera on a tripod to get a steady shot and then plugged the camera into the computer with USB. The camera *mounts* (shows up on the computer screen) as another drive. Then I simply dragged the file with my mouse from the camera to the desktop and then into Vegas Movie Studio so I could edit it. (You'll learn more about Vegas Movie Studio later in this chapter.)

Here's motion video called "Mr. Peep's Wild Ride" done with the same Canon still camera. Since it's in a moving car, it doesn't look as good:

www.youtube.com/watch?v=0zY2uL55Zss (URL 3.10)

But for close-up stuff, like "Kitty Lickin' Good," and for vlogging, shooting video on a still camera is OK, and really cheap.

Some cheap still cameras have built-in flash memory but not much of it. They shoot only about 3 minutes of grainy video. You actually could shoot something on these and put it on YouTube, and if you have nothing else to work with and no money for a better camera, I say go ahead and get started with whatever you have. But with better still cameras (like my Canon PowerShot), you can shoot video that doesn't look as good as mini-DV but looks OK.

The better still cameras use removable memory cards, so you can buy a larger-capacity memory card (make sure you get the right format; there are a few of them) and shoot longer clips. My Canon PowerShot A460 cost only $110 and came with a 16 MB memory stick. But with the $10 512 MB SD card I bought on Amazon, it can shoot about 20 minutes of video.

Shooting video for YouTube on a still camera can work well, especially if you record the audio separately. The microphones on most still cameras suck, and the encoding rate is low. It's a little more work, but it's worth it if you're using a really cheap still camera or a cell phone camera and want to have a chance of going viral on YouTube.

PRO TIP

If you want to know what photo quality is like for a particular still camera you are thinking of buying, search for the make and model on Flickr.com, and you'll get many examples of photos taken with that model. The look of the still shots a camera shoots gives a rough idea of how video shot on that camera will look.

Recording sound separately

You can record sound separately in several ways. You can plug a lavaliere microphone into your computer and record directly into your video-editing program and sync up with the video later. (You'll learn more about this in "The Importance of Audio" section, later in this chapter.) Or you can record your audio on a dedicated digital recorder and later import the audio and video separately and sync it in your editing program. I'm fond of this method and record on my Zoom H2.

The Zoom H2 is an amazing portable digital recorder. I call it the "studio on a stick." (It comes with a detachable handle that sort of makes it look like an ice cream bar, and with an alternate stand, a little tripod for table use.)

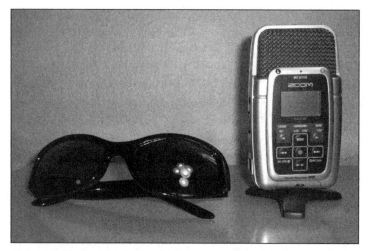

Figure 3-7. The lovely and powerful Zoom H2 ready to record.

The Zoon H2 (Figure 3-7) is $179 and available from:

www.zzounds.com/item--ZOMH2 (URL 3.11)

The Zoom H2 is about the size of an electric razor (and looks rather like one) and records CD-quality sound (44.1 Hz 16-bit stereo WAV files) to an SD card. (Again, if you want to record anything of length, you'll need to get a bigger card than the one that ships with it. I got a 4 GB card for 40 bucks and can record six hours of CD-quality sound on the amazing built-in condenser mics.) You can record stunning sound on this thing using the built-in mics and then sync it later to your video.

Here's a blog post I did with a review of the mighty H2:

www.stinkfight.com/2007/08/31/my-review-of-the-new-zoom-h2-portable-digital-recorder/ (URL 3.12)

Here's a 55-second video I made that shows the difference between the sound of using a webcam's built-in mic and using sync sound from the H2:

www.youtube.com/watch?v=QOcrwDFZGOo (URL 3.13)

The H2 can also record much longer program material on the same drive if you change the settings to record MP3 instead of WAV files, but you don't want to do that. MP3s are a compressed, degraded format. WAV (and AIFF) are uncompressed. You want the highest-quality sound you can have, especially since YouTube degrades the audio even further when it compresses it to MP3 (within the Flash video interface it uses). MP3s made from MP3s sound horrible.

You don't need to use the MP3 setting on the H2 anyway, because you're shooting short films for YouTube. The 512 MB card that ships with the H2 will record only about 5 minutes of WAV audio, so you might want to invest in a 1 or 2 GB card. You want to be able to record more than one take of your show, because you will have more to work with when you edit that way.

When I do a vlog, I usually have my wife hold the camera and put the H2 on a mic stand. You could also tape the handle to a broom handle and stick it in a milk crate if you don't have a mic stand. Make do with what you have.

If I'm alone, I put the camera on a tripod or a mic stand. If someone is in the house with me, I'll ask her to hold the camera for me. Having a person hold a camera, even if they hold it as still as they can, will add a bit of nonstatic feel, of movement, to the shoot. This can be useful with shooting a person talking, especially if the "human tripod" doesn't overdo it.

Here's another video that is a good demonstration of handheld camera work:

www.youtube.com/watch?v=_kxT8S-Er0g (URL 3.14)

Note Being good at one thing in life, especially art skills, will make you better at everything else. Acting will make you a better director, and editing audio and video will make you a better writer; your skill in one art is transferable to other art forms. That's one of the secrets to life. The other secret to life, the main secret to life, is that it's "better to give than to receive." Seriously. That's the secret to life. And you thought this was just a tech book. Ha! Tech books are for nerds. This is a book for entry-level geniuses. Keep reading. You'll see.

You can get mic stands (and a lot of other great musical gear) cheap here:

www.musiciansfriend.com (URL 3.15)

One mic I used in that video is probably overkill for what you'll need to do, but I'm a music snob. The mic is a condenser MXL M3-B, a prototype of the MXL M3. The MXL M3-B was never released, and I'm honored to own one. A cool guy from the Marshall guitar amp company gave it to me. (They make MXL mics.)

Figure 3-8. My MXL M3-B condenser mic set up for recording separate audio while shooting video.

Large condenser mics like this (Figure 3-8) require an external power source, are very fragile, and are very susceptible to moisture, smoke, and dust. I don't recommend them for entry-level YouTube video making, but if you have one and know how to use it, your videos can only sound better.

Keep in mind that these mics (and all mics) are very sensitive and pick up everything, from the air conditioner in the next room to the truck out on the street. So, they're kind of overkill unless you're an audio expert. But with any mic, any video recording that includes audio, you should try to minimize background noise as much as possible, because it's nearly impossible to completely remove it in post, and removing it affects the overall sound negatively. So, before you do a video shoot of any kind, even a vlog, turn off the air conditioner, get as far away as possible from your computer fan, have your roommate turn down her music, and shut the windows. Your audience will thank you. Even if they don't know what they're hearing, they'll know it sounds better and like the video more, your video will get more views, you'll get more subscribers, and you'll have a better chance at going viral.

PRO TIP

Mini-DV cameras have something called *timecode*. For basic YouTubing, you don't really need to understand this, except to know that in advanced video editing (which Alan will touch on later), it helps keep things absolutely mathematically synchronized (synced) from start to finish. If you have two pieces of hour-long mini-DV video that are shot of the same scene at the same time, you should be able to sync the beginning of one to the beginning of the other, and they should be perfectly synced, to the frame, at the end.

Webcams, still cameras, and cell phones do not have timecode. So if you shoot even 10 minutes (the YouTube max) with one of these cameras, the audio you record separately is not going to sync perfectly. At the end of 10 minutes, people's mouths will be a little ahead or a little behind the sound of them speaking. It will drift about a quarter second per 10 minutes, which is noticeable, particularly on close-ups.

I've figured out a workaround for this, which is to sync the audio in the middle, not at the beginning, so it averages out the drift to less at every point from the beginning to the end. On a video less than 5 minutes, one that doesn't have any extreme close-ups of the subject's mouth, this will make drift virtually unnoticeable. The way I sync externally recorded video is to add an extra audio track in the editing program, drag the external audio into the program, and then move it left and right until there is no echo with the audio recorded by the camera. Then I mute the original camera audio. You may need to zoom in or out in the time view in your program to do this.

Here's a vlog I did, in which I recorded the sound separately on my H2 recorder and synced it with the video later. You can see the H2; it's in frame, on the right. It sounds really good, except it picked up the sound of a fan in the next room. I forgot to close the door. Let this be a lesson to ya!

www.youtube.com/watch?v=vbZdcEoXeGQ (URL 3.17)

The Importance of Audio

Audio, believe it or not, is the most important technical aspect of motion art. The audio is even more important than the video. If you take two clips, the first with great video and crappy audio and the second with shaky video but great audio, and show them to a random selection of people, most people will enjoy the shaky one with great audio more. And if you take two clips with equally good video and one has better audio, people will like the one with better audio more. Even if they don't know what great audio is or how to describe it, they will appreciate the difference. Great audio just really pumps up the professionalism of how video is perceived.

My biggest recommendation for filmmaking (and it's one that's mostly ignored on YouTube but not on the viral videos) is this: Use an external microphone. The microphones built into most consumer cameras suck. Moreover, you'll pick up the sound of the camera's motor with the onboard mic. And the mic will be too far away from the subject's mouth to get great sound.

Again, your Audio-Technica ATR35S lavaliere mic comes to the rescue.

The sound recorded by most still cameras is horrible. You cannot use the sound from a still camera and expect to compete on YouTube. And you can't plug a lavaliere mic into them. They don't have an input jack.

For getting great sound with a cheap still camera, I recommend using the lavaliere mic to record the audio separately, directly into your computer, into Vegas Movie Studio.

www.sonycreativesoftware.com/moviestudio (URL 3.16)

After you import your video into your editing program, you then import your audio, sync the good audio with the crappy camera audio in the timeline, then mute the crappy camera audio before you edit, and render. (You'll learn more editing in the "Editing" section, later in this chapter.)

Background Music

Background music contributes immensely to the feel of a video. Your choice of music can make or break the video.

You have to be careful with copyrights when using music by other people. I recommend getting music from the Podsafe Music Network to use in your videos.

http://music.podshow.com/ (URL 3.19)

It's free to use, with attribution. Putting a link to the composer in the "More Info" section of your video's YouTube page should satisfy this requirement, but read the terms of service on the Podsafe site, and on the individual music, because this can change.

The mood is not the only thing that's important with your music choice. It's also important to choose music that doesn't compete with talking and use it at a volume that supports the story, rather than distracts from it.

The "Background Music" sidebar is reprinted from something I wrote for the O'Reilly Digital Media website, *www.oreillynet.com/pub/au/3220* (URL 3.20). It has some great info on background music choices.

BACKGROUND MUSIC

Some types of music work better than others under speech. Music with *any* vocal sounds (singing, rapping, talking, grunting—anything produced by a human voice) is going to distract from talking on top of it. Also, music with a lot of dynamic range (variation between very loud and very soft) is not usually a good choice. Nor is music that has any samples or percussion that sound like noise (in other words, material that has a "spray" or "water falling" or "wind" sound). And music with very tinny or bright-sounding percussion, especially cymbals (real or sampled), will usually compete with any talking on top of it.

A lot of techno and industrial is a bad choice for *beds* (the industry term for background music) because of the noise factor and/or bright percussion.

Basically, there's a big difference between good music and good *background* music. A lot of music I love to listen to makes horrible background music to talk over. Conversely, a lot of music I wouldn't listen to on its own makes great background music.

I find the use of distracting beds a mistake that's often made in independent filmmaking (and occasionally in pro filmmaking). Inexperienced directors will use music with vocals that sums up the mood of the film and then have people talk over it, and it's hard to hear the people talking. For instance, a punk-rock track with shouted vocals or a hip-hop track with rapping on it is great to set the mood for a film. (Or a podcast or any audio media. I look at podcasts as "little movies without visuals" and approach my production with this in mind.)

But music with singing is not great background music for scenes in a drama where characters are talking or in a documentary film under interviewees speaking. It's best to save vocal music for montages where there's no speaking or use it to set a mood before or after actors (or interviewees) speak.

I don't listen to hip-hop much in the course of my week, but I tend to use it (without the rapping) in films I make and under talking in podcasts. Hip-hop is *made* to talk over, ya know? Rapping is talking, albeit rhythmic talking.

Trip-hop (mellower, spacey music made with similar production techniques to hip-hop) makes even better background music under talking. It has the level dynamics of hip-hop but is a little mellower and less obtrusive.

It's really easy to make great background music in Sony Acid, and Sony even offers a free version, Sony Acid Xpress. Mac users should check out GarageBand.

Here's a YouTube video I did that demonstrates types of background music that work and ones that don't:

www.youtube.com/watch?v=NAnYj-iNrPQ (URL 3.18)

Levelator

The Levelator (Figure 3-9) is an audio secret weapon that is very useful to filmmakers. It's made and given away free by a very cool company called the Conversations Network. Originally made for podcasters, it's great. I use it for film. The Levelator is free software for Windows, Mac, or Linux that does only one thing and does it very well. It takes files of spoken audio and evens out the sound. It makes stuff that's too quiet a little louder and stuff that's too loud a little quieter.

It works only on speech. It won't improve the sound of speech mixed with music or other sounds; in fact, it will make that sound worse. But you want to have more control over your audio, so you should record the talking first and add the music later anyway. That just makes for better overall sound.

The Levelator is drag and drop, with no controls. You can't adjust anything. You just open it and drag a WAV or AIFF audio file onto it. It starts processing the audio and gives you a status bar showing the percent of progress. When the bar reaches 100 percent and goes away, your media is done *cooking*, and there's a new file with ".output" added at the end of the filename. That's the one you want to use.

www.conversationsnetwork.org/levelator/ (URL 3.21)

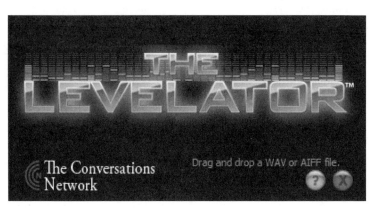

Figure 3-9. The simple interface of the Levelator.

Video from stills

Stills can be used to great effect in video. You can drag them in and use them with video, or you can make video entirely out of stills. Used sparingly, you can stand out by using stills to spice up an otherwise static vlog or other video. With a book, you can show the cover for a few seconds while you talk about it. Use your imagination to think of unique ways to use stills, and you'll be ahead of what most people are doing.

Here's a three-part article, "Put Your Photos on TV," that I wrote about making video from stills. (Part 3 was cowritten by David Battino.)

www.oreillynet.com/pub/a/oreilly/digitalmedia/2008/02/28/put-your-photos-on-tv-pt1.html (URL 3.22)

www.oreillynet.com/pub/a/oreilly/digitalmedia/2008/03/06/put-your-photos-on-tv-pt2.html (URL 3.23)

www.oreillynet.com/pub/a/oreilly/digitalmedia/2008/04/17/hi-res-youtube-hacks.html (URL 3.24)

Cell Phone Cameras to Capture the Breaking Story

Most cell phones shoot video these days. Some shoot really good-looking video. Any of it can be converted, edited, and uploaded to YouTube.

Cell phones are often used to capture breaking news—events like fights, hangings, unrest, or celebrities doing things that someone, for some reason, finds interesting.

But you don't have to use a cell phone for breaking news only. If your cell phone is the only camera you have, you can use it as your primary video capture source. No matter what type of camera you use, everything you learn making films will help you later making better films. And you'll also get used to navigating and promoting on YouTube. So if you don't have money for a better camera but have a cell phone that shoots video, start now anyway.

My friend George Earth, does some amazing low-fi cell phone movies of his series "Cheap Comix" on YouTube.

www.youtube.com/user/candymachinegeorge (URL 3.25)

www.georgeearth.com (URL 3.26)

George writes the stories out as comic strips, which you can download here:

www.stinkfight.com/2007/09/11/cheap-comix-2-up-for-download-now/ (URL 3.27)

Then he makes cardboard cutouts of his characters and films them with his cell phone, while doing the voices live. He even makes little sets and towns for them. He is, by his own admission, not very good at drawing, but the dialogue and execution make them particularly compelling. Here's one of them, called "What's Performance Art?"

www.youtube.com/watch?v=XTPLHXFKsQg (URL 3.28)

Cell phones usually use proprietary formats and can take a lot more fluffing than mini-DV to get the video into your editing program. But this can be great if you shoot a breaking story on your cell phone because it's the only available camera, or if you have a cell phone that shoots video and can't currently afford another camera.

PRO TIP

Check out this video:

www.youtube.com/watch?v=rlNTqlEc9Nk (URL 3.29)

George's friend Anarchy Jordan transferred the proprietary cell phone video format into the editable .mov format. He also did the moving mouth animation. Anarchy Jordan writes: "The cell phone records video in the 3gp format, which I had to convert to .mov format using FFmpeg (free to download).

http://ffmpeg.mplayerhq.hu/ (URL 3.30)

"I used a free open source 2D animation software program called Pencil to make the moving mouth.

www.les-stooges.org/pascal/pencil/ (URL 3.31)

"With that program, you can draw little pictures and set them up as frames in a movie file. Then I took that and put it on the little cutout guy's face in Final Cut and made it follow the motion, using Chromakey in my editing program to make it transparent around the edges and the motion options in my editing program to get it to follow the jerky movement of George's hand as he filmed his cutout guys with his cell phone camera.

"The object (the mouth) can be made transparent wherever it's made up of a certain color. For instance, in this example, the 'moving mouth' was originally just a little drawing on a white background. Chromakey takes out all the white and makes it transparent, leaving only the black. Kinda like green screen for different colors."

You can learn more about Anarchy Jordan's view of the world here:

http://situationist.gq.nu (URL 3.32)

What Not to Shoot and Why

YouTube doesn't need another video of a tin can in a microwave. And it certainly doesn't need another video of some guy wrecking his family jewels on a railing while falling off a skateboard. Sure, this stuff gets lots of hits. But it's not going to help you bring people back again and again.

After you've made and uploaded your first few videos of…well, anything, and gotten over the tiny technical hurdles of making stuff look and sound good, then get original. Figure out how to bring something original out of your own twisted little mind and make it fly. Don't worry too much at first about writing for an audience; that comes later, after you've done more experimentation. Just try to do something that reflects you.

Look on YouTube, and see what's been done to death. Then do anything else. The only reason I can see to do something that's been done to death is to parody it. So, find something unique or that expresses a unique viewpoint, and do a drama, comedy, or vlog about that.

Location

Locations for YouTube videos can be anywhere. Many are shot in people's bedrooms or yards. But it can be more fun to get "out in the world" and "away from the nest" to experience and film more of what the world has to give. You want your videos to be special, so don't limit them to your room.

When Hollywood film crews shoot in public, they have to pay for permits. Some cities require this even for independent filmmaking. (Santa Monica, California, comes to mind.) I would never recommend you break the law, but I will say this: I've never gotten a permit for anything I've filmed. And I've never gotten in trouble. And there's a good chance that anything filmed for YouTube is so far "below the radar" that it probably isn't an issue.

Basically, I just film anywhere. In stores, in churches, in parking lots, in people's front yards. I'm respectful, quick, and if asked to leave, I do, and do so politely.

Lighting

Mini-DV cameras, still cameras, and webcams are pretty forgiving on a wide range of light levels, much more so than most film movie cameras. Even so, it can't hurt to brighten up indoor shoots with some lights. And you don't need fancy movie lights. Little $10 halogen lights from Target or any place similar are more than enough.

Lights will bring out colors more. You can experiment with light placement to cast interesting shadows. People look evil when lit from below and angelic when light from above. Also try lighting a face in a dark room from one side only so one side is lit and one side is dark or almost dark. It's a nifty effect. Colored lights can also make cool effects. You can get sheets of colored gel at any art-supply store and hold them over lights. (Don't do this with halogen lights, though, which get too hot and will melt or burn the gels. And be careful with halogen lights in general. They get really hot. And if the quartz glass covering over the lighting element ever cracks, throw the light away, because that quartz filters out the part of the light that can damage your eyes.)

I bought some cheap halogen lights at Target. The vlog I did of my "50 things" had these lights shining on my face.

Props

Ikea, Staples, The Home Depot, and the 99-cent store are great places to buy props, gear, sets, and even lights. Walk around, look at things, and let your imagination run. Bring your camera. I can think of no reason not to write and shoot a story on the fly, even in the store, if you aren't obnoxious and if you stop if they ask.

I often spontaneously think of short plots when I have a handheld fan or a power drill and I'm looking at sheet rock. Or petting a giant calculator. Or palming a fork with fuzzy oven mitts. (See Figure 3-10 and Figure 3-11.)

Figure 3-10. Improvised office skit using found props at Staples.

Figure 3-11. Improvised monster skit using found props at Bed Bath & Beyond.

Computers

In my mind, the war between PC and Mac is over. They're both so similar now, from years of PCs trying to look like Macs and Macs now running on Intel chips, that they're the same. (Except Macs look sexier, they cost a lot more, and there's not as much software for them.)

Alan and I use PCs because it's what we started on and because, since they're cheaper, we've seen no reason to switch.

Another good reason to stick with PCs is that Vegas Pro and Vegas Movie Studio run only on PCs. Here are the system requirements for Vegas Pro:

- Microsoft® Windows® XP SP2 or Windows Vista™
- 1 GHz processor
- 200 MB hard-disk space for program installation
- 1 GB RAM

Most new PCs have all of this, but check when buying a computer.

Here are the system requirements for Vegas Movie Studio:

- Microsoft® Windows® XP SP 2 or Windows Vista™
- 800 MHz processor
- 200 MB hard-disk space for program installation
- 256 MB RAM

If you already have a Mac, you can use iMovie, which now comes on all Macs. If you have a PC and cannot afford Vegas Movie Studio, PCs come with Windows Movie Maker.

The interfaces for most editing programs are pretty similar, with File/Open/Edit; some sort of cut, chop, and move function; and a timeline for your media, with separate tracks for video and audio.

PRO TIP

You can get a lot more bang for your buck—more processing speed, memory, and hard drive space—if you buy a tower than if you buy a laptop. You can get the same specs in a tower for a third of what a similar setup costs in a laptop.

I recommend NewEgg for buying computers, computer memory, and peripherals.

www.newegg.com/ (URL 3.33)

The site is great, and its prices rock. It tends to ship really quickly too, especially if you live on the West Coast, usually within a day or two. Once I bought a printer from NewEgg with standard (not expedited) shipping and received the printer via UPS in less than 24 hours. That's kind of mind blowing.

If you're really on a budget and know something about setting up and fixing computers, another option, sort of a last resort really, might be this site:

www.propertyroom.com/ (URL 3.34)

It has online auctions for computers (and other things, including land) seized by the police. The computers have no software or operating system; they've been wiped clean. Some are even listed as having had the hard drives removed. And some have to be picked up in person. But people can get decent computers on there for $30 to $100; put in a $40 hard drive (from NewEgg perhaps?) and an operating system, and be working way cheaper than you can get a computer anywhere else.

(I wouldn't buy land on that site; I'd be afraid that the former owner would get out of prison and come back to claim it, but laptops should be OK.)

Editing

Now that you've gotten your brilliance into your camera, you need to get it out of your camera, chop out the parts that don't work, and distill it down to its happy, perfect goodness. So, take a break, get up, stretch, walk around, have a soda, and we'll meet you back here in a minute.

Free, Cheap, and Pro Audio and Video Software

Alan is going to take over for the rest of the chapter to cover basic editing on the inexpensive Sony editing program, Vegas Movie Studio ($74.95).

Later Alan will cover some advanced editing techniques with Vegas Pro in Chapter 8.

Vegas Pro costs $549 or $305 with a student ID from Academic Superstore.

www.academicsuperstore.com/ (URL 3.35)

Vegas Movie Studio looks and works much like Vegas Pro, without all the bells and whistles, and it does everything you would need for basic editing of vlogs and short videos. It can also export to WMV. And that's a good thing, because the WMV format optimizes well for YouTube.

Basic Editing

Alan Lastufka

Editing neither is as overwhelming as it sounds nor is it as overwhelming as it looks when you first open your chosen piece of software. Most nonlinear editing (NLE) software contains multiple windows or docks and numerous, colorful, but unlabeled icons. However, given the content you'll most likely be working with and the small number of edits and transitions you'll need, you'll be editing like the pros in no time.

Why Bother Editing?

As you've no doubt watched a few YouTube videos before picking up this book, you may have noticed that some YouTubers don't edit their videos at all. You may ask, why should I bother? Well, some videos require no editing. A three-minute vlog, recorded in one take with little or no mistakes, requires no editing at all, unless you want to remove the "ums" and "uhs." Other videos may require numerous edits for time restrictions, content, or pacing. Editing involves chipping away at all of the material you have, until only the best footage remains.

Style is another important factor when discussing why you should edit your videos. Most of the "popular" vloggers have perfected the jump cut and use it effectively for funny or quirky transitions. I'd estimate that 90 percent of the edits you'd see on a typical YouTube viewing day are jump cuts. Jump cuts will be defined in the following section.

The final factor in why you may need to edit instead of just uploading a raw take is whether you plan to add any music, text, or collaboration clips into your video. While expensive equipment exists to do this in real time, it is out of the budget for almost all vloggers, so these elements must be added after you record, in postproduction.

Basic NLE Editing

For the basic editing demonstrations in this book, I will be using Sony Vegas Movie Studio. Movie Studio is a very affordable yet powerful NLE aimed at the YouTube crowd of video makers (see Figure 3-12).

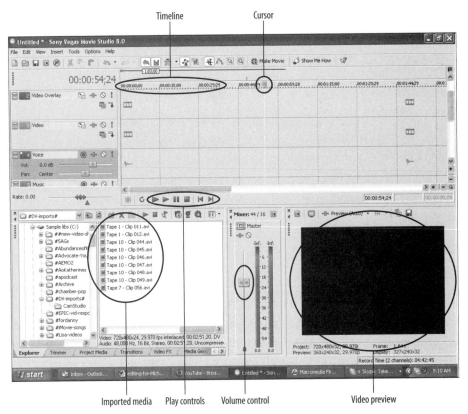

Figure 3-12. Default screen layout for new projects in Sony Vegas Movie Studio.

You could always spend more, or less, depending on your means and needs, but I've found Movie Studio to be a good price-to-feature value. While some of the keyboard commands in Movie Studio may differ from your editing software, the techniques and fundamentals will remain across programs and platforms.

Capturing Video from Your Camera

Before you can begin editing your footage, you must "capture" it from your video camera. Some cameras come with video-capturing software prebundled, but these programs are usually limiting or capture in a proprietary format (a media type that you can't easily import into other programs). It's better to use the capturing feature found within your NLE.

To capture video, make sure your camera is correctly connected to your computer via FireWire or USB, as shown in Figure 3-13 and Figure 3-14 (or Figure 3-13 and Figure 3-15, if you're on a laptop). FireWire ports may be found on the front or back of your computer if you have a tower computer and are usually on the side if you have a laptop. If your laptop doesn't have FireWire ports, you'll need a FireWire card, as shown in Figure 3-16. Also, keep in mind that your camera may require you to use a certain setting when transferring or capturing video; please consult your camera's owner's manual if you run into any trouble.

Figure 3-13. FireWire cable connected to the camera.

FireWire cable   FireWire ports

Figure 3-14. FireWire cable connected to the port on the back of a tower computer.

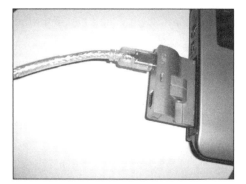

Figure 3-15. FireWire cable into FireWire card into laptop.

Then, from the File menu, choose Capture Video, as shown in Figure 3-16. This opens a new window or dialog box, as shown in Figure 3-17 and Figure 3-18.

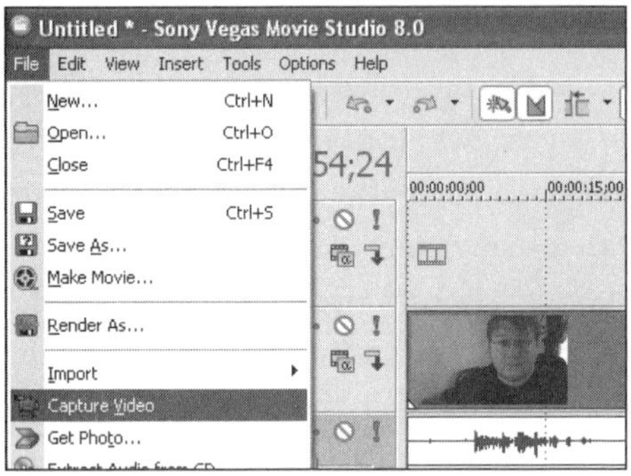

Figure 3-16. From the File menu, choose Capture Video.

Figure 3-17. A typical Video Capture dialog box before a device is connected.

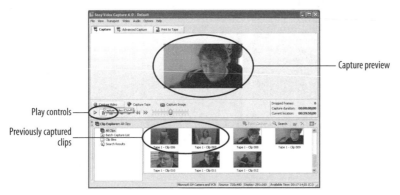

Figure 3-18. A typical Video Capture dialog box after a device is connected.

This video-capturing dialog box may contain simple Play, Rewind, and Fast-Forward functions to find the spot from which you'd like to begin rendering (and giving yourself a bit of *preroll*—that is, capturing a few seconds before the spot at which you want your video to begin—might help you while editing). Once you find the spot, hit the Capture button.

If recording to DV tape (the most popular method), you will have to transfer your video in real time. This means you can watch your video playback as it's captured and stop it when the scene ends. If you're using some sort of memory card or flash drive, you can transfer your entire video to your hard drive much more quickly.

After the capturing is complete, you are ready to view and edit your footage.

Importing Media

You may want to use media beyond what you've captured with your camera in your videos. Perhaps your friend's band has allowed you to use their music, you make your own music, or you've designed a really neat title card in Adobe Photoshop. Either way, you're going to have to import that media as well. Thankfully, because you'll be doing this a lot, most programs have made this function just a one-click process.

In Vegas Movie Studio, you simply click the Import Media button, found right above the Project Media window, as shown in Figure 3-19.

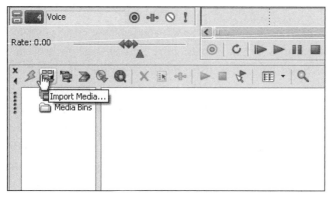

Figure 3-19. Import Media button.

The Import Media button will allow you to browse your computer for any file that Vegas Movie Studio can handle, from still images to animated GIFs to video files and audio, including AIFFs, WAVs, and MP3s.

Once imported into Vegas Movie Studio, you can simply drag and drop your files into the timeline just like the footage you captured earlier in this chapter.

Editing Choices

Editing has very few hard and fast rules. You certainly can get away with using various transitions like *wipes* (where one scene slides over another, bumping it off the screen) or *cross-dissolves* (where one scene fades into the next). But these transitions get in the way and usually distract from the material you're presenting. In other words, people notice these types of edits more than the content. People new to editing tend to do fancy transitions because they can, not because they should. It's usually a mistake.

The best edits are seamless. This doesn't mean the trained eye can't see them; they are, after all, cuts that remove frames from the continuous motion, but if done with care, most viewers won't even notice a new camera angle or position. You can achieve this in a variety of ways, some of which are outlined here:

- *Jump cut*: A jump cut is a transition that simply "jumps" from one scene to the next. In most vlogs, you will typically see the YouTuber jump around in the frame as they cut from one sentence or paragraph to the next. This may be used to cut out mistakes to shorten the overall video length by cutting out pauses or for comedic effect, such as cutting to a new standing position after the punch line of each joke.

- *L-cut*: An L-cut is used a lot in Hollywood blockbusters during conversations, especially phone call conversations. If we were to jump cut from one piece of dialogue to the next, the conversation feels very distant and separated, like two people delivering lines. However, if you overlap the video and audio tracks (which visually form an L-like shape in your NLE) during a conversation, your viewers can now see reactions from the person not speaking, while hearing the voice of the speaker. These reactionary shots can be short and hold only for a moment or two before you cut to the matching video, but they will really sell two separate shots as one continuous conversation.

- *Other cuts*: Depending on which NLE software you've chosen, you may have an array of other premade transitions available to you, such as the wipes or dissolves I mentioned earlier, along with flashes, pushes, and other reveals (which are fancy forms of transitions). Use these transitions sparingly, though; overusing these effects is the mark of an amateur, and they really distract from your material.

OK, so now that you know what's available, let's go ahead and look at how to perform the most common edits.

The jump cut

The simplest, most effective, and most used of the three edits discussed earlier, is the jump cut. To perform a jump cut, you are simply going to remove a given number of frames between two points in your video. Perhaps you stumbled over your words and started a line of dialogue over, or perhaps you received a phone call and had to pause your recording for a moment. Either way, to perform a jump cut, place your cursor at the point where you want to make your first cut and, with the track selected, hit the S key on your keyboard, as demonstrated in Figure 3-20 and Figure 3-21.

Be sure to first select the piece of footage you'd like to cut. If more than one file is beneath the cursor and neither is selected, pressing S on your keyboard will cut/slice both pieces of footage.

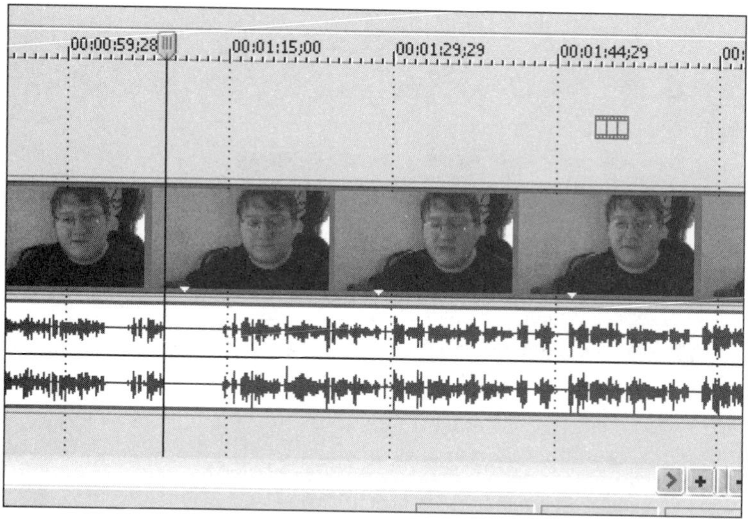

Figure 3-20. First, position the cursor where you'd like the cut to take place.

Figure 3-21. Second, hit the S key to perform the cut or slice.

Now, move the cursor forward through your footage until you find the start of your next scene/sentence/topic. Place the cursor right where you'd like the next piece to come in, and again hit the S key on your keyboard to cut/slice it, as demonstrated in Figure 3-22 and Figure 3-23.

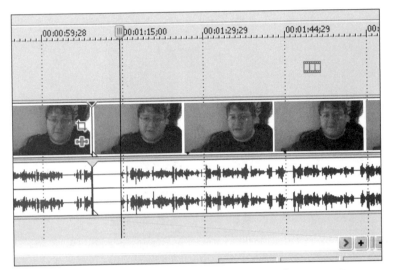

Figure 3-22. Third, position the cursor just before the start of your next piece.

Figure 3-23. Fourth, hit the S key once more to perform the cut/slice.

Now we are going to remove the unwanted frames between the first and second cuts. To delete these frames, simply click anywhere between the two cuts, and hit the Delete key on your keyboard, as shown in Figure 3-24.

Figure 3-24. Highlight and delete the unwanted frames between the two cuts.

Now, we are going to use the mouse to slide the remaining frames over in the timeline to butt up against each other, as shown in Figure 3-25.

Figure 3-25. Slide the remaining frames over into position.

The L-cut

L-cuts are performed in a way similar to jump cuts, with one additional step: you must separate, or ungroup, the video from its audio track.

When you capture or import a new video file, its audio track is automatically grouped to it, so, as you move around the video frames, the audio remains in perfect sync. However, when we perform an L-cut, we want to actually remove frames, without removing the audio.

To ungroup any tracks in Sony Movie Studio, right-click either of the tracks (which will open a contextual menu) and choose Group > Clear, as shown in Figure 3-26.

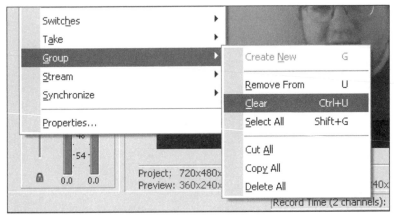

Figure 3-26. To ungroup, simply right-click either track and choose Group > Clear.

Now you can freely move (or cut/slice) the video independent of the audio. Be careful to not slide the video out of sync with your audio. Use the ungrouping feature only to cut the edges of the video and audio independently, and then regroup the video and audio tracks.

To regroup the video and audio tracks, repeat the process of selecting both tracks by Shift-clicking them and then right-clicking either of the tracks. Then choose Group > Create New.

Figure 3-27 shows what a typical L-cut edit should look like. You'll notice that the audio from our second clip extends beneath the video of our first clip, so, as you play back the edit, you would hear the audio of the second clip play while still seeing the video from the first clip. Then, the second clip would appear as a jump cut, and the audio would then be in sync with the second clip.

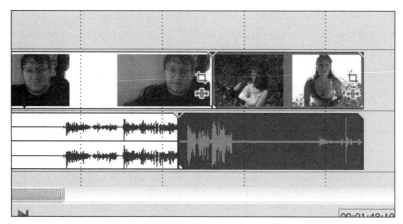

Figure 3-27. Sample of an L-cut in Vegas Movie Studio.

 L-cuts are also sometimes referred to as *J-cuts*, depending on which way the audio and video extends. To simplify things here, we'll refer to them only as an L-cuts.

Other cuts

Depending on your choice of NLE software, you may also have additional transition options available. These could include fancy wipes, flash transitions, and numerous other playful elements. I won't go into detail here about using these effects because the application and their functions will vary greatly across programs, but searching for the word *transitions* in your user's guide or manual will provide you with all the information you'll need for applying these effects.

Again, as I cautioned earlier, use these effects sparingly. They are exciting at first but can be very overused, which will make you look amateurish.

Rendering

Once your video is edited the way you'd like, you must take all the pieces and put them together into one file. This process is called *rendering*. Picture rendering as "cooking" the ingredients into the finished cake. Rendering essentially takes all the clips of audio and all the clips of videos, including transitions, effects, and other elements, and it squishes them into one, *uneditable* file. I stress the "uneditable" because if you throw away your raw footage files, you can't easily use your rendered file to extract individual pieces of footage or audio.

Rendering is the final step in the editing process before you upload. When you render, you will also be compressing your files using a *codec*. You don't need to understand codecs to understand why they are useful. Various codecs exist to compress your footage into a useable-sized file. Uncompressed video is huge…very huge. Uploading uncompressed video would take hours or days depending on your connection. So when we render, we choose a compression format, lose a little bit of quality, but save a lot of hard drive space.

To render your footage, first set beginning and ending markers in your NLE by clicking and dragging your cursor directly above the timeline, as shown in Figure 3-28.

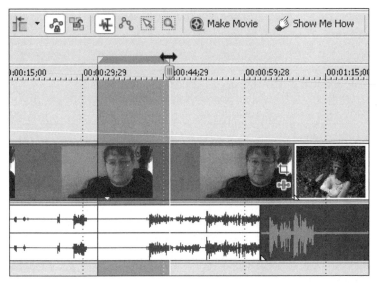

Figure 3-28. Click and drag above the timeline to set your beginning and ending markers.

This tells the program which part of the project is to be rendered. Again, this will vary for each NLE, but you should have markers similar to those shown in Figure 3-29. You can also hit the I key for the *in point*, or start, and the O key for the *out point*, or end. This works on many other NLE programs, including Vegas Pro.

Figure 3-29. Beginning and ending markers set the start and stop points of your render.

From the File menu, choose Render As. (This may also be labeled as Export, depending on which editing software you're using.) This will open a dialog box with various rendering options. From this dialog box you can choose the file format you'd like to render to (.wmv, .mov, and .avi are the most popular choices), and you can set the compression settings as shown in Figure 3-30.

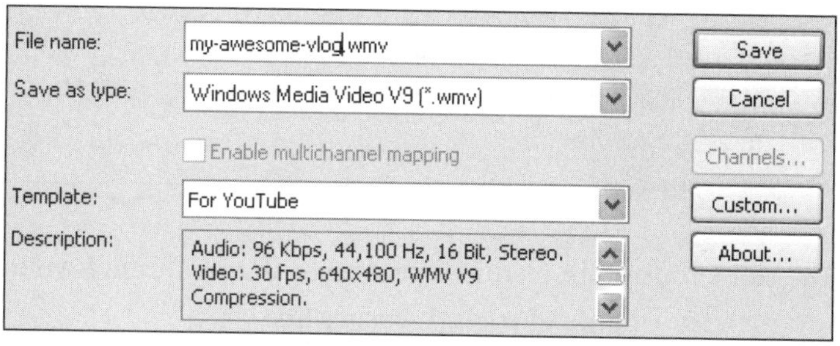

File name:	my-awesome-vlog.wmv	⌄	Save
Save as type:	Windows Media Video V9 (*.wmv)	⌄	Cancel
	☐ Enable multichannel mapping		Channels...
Template:	For YouTube	⌄	Custom...
Description:	Audio: 96 Kbps, 44,100 Hz, 16 Bit, Stereo. Video: 30 fps, 640x480, WMV V9 Compression.		About...

Figure 3-30. Choose your settings before rendering.

I work on a PC and therefore use .wmv for my file format. As for compression settings, I've found that the following settings work well for me in a quality to file size ratio. You can adjust these settings after clicking the Custom button, as shown in Figure 3-30.

Video Rendering Quality: Good
Video Mode: Quality VBR
Video Quality: 90%

You may have additional fine-tuning options available, but I'd leave most of those set to the default setting unless you know what you're doing.

When you're finished, add a title to your file, and click Save, which will begin the rendering process. Rendering can take anywhere from 2 to 10 times the amount of running time of the video; it all depends on your computer's processor, the number of effects and transitions, and your computer's free memory. For instance, a 4-minute video can take between 8 and 40 minutes to render.

The resulting rendered file should now be about 10 MB in size for every 1 minute of video (a 4-minute video would total roughly 40 MBs). This rendered file is the one you should be uploading to YouTube. As mentioned earlier, you should hang on to your larger, uncompressed, raw files if you believe you will ever need to reedit or reuse any video or audio elements in a future project. To save space on your hard drive, you can back them up to data DVD or get an external FireWire drive to back up many projects.

PRO TIP

As you learn more about video compression and what happens during the rendering process, don't be afraid to experiment with various rendering settings by clicking the Custom or Advanced tabs in the rendering dialog box. Many settings can greatly affect file size and quality. Once you have the settings tweaked to your liking, save your rendering settings as a template for future use.

Your editing program has numerous other features. Be sure to thumb through your owner's manual, or check out a few online tutorials. Websites like

www.creativecow.net (URL 3.36)

have wonderful, free video tutorials that cover Vegas, AfterEffects, and many other popular video-editing software. As with any hobby, the more you do it, the better you become.

With that in mind, let's move on to Chapter 4, where you'll learn how to set up your YouTube profile so you can begin uploading videos.

4

Creating Your Very Own Channel

Alan Lastufka

All About Your Channel

What most other social networking sites call your profile, YouTube calls your *channel*, as in "TV channel." Picture your YouTube channel as a TV transmitter that reaches the world from your room. When I refer to your *channel* in this chapter, I am referring to your profile and/or your account on YouTube. Your channel is your home on YouTube; it is a page that resides at the following address:

http://www.youtube.com/user/YOURUSERNAME

Your channel page contains a description, which is information you enter about yourself, such as your likes and dislikes, favorite books or films, and other items. Your channel page also displays any videos you upload, along with the videos you save as your favorites. I will explain how this works in detail throughout this chapter; just know that your channel is where almost everything you do on YouTube is done.

Registering Your Account

Registering a YouTube account is simple. Although many of YouTube's functions are available to unregistered guests, such as video searches and watching videos, registering an account offers many benefits. As mentioned, your YouTube account is your home on the site. Without an account, you are simply a spectator. This book isn't for spectators; this book is for those who want to be seen. *Don't just watch the media. Be the media.*

Registering an account allows you to leave comments and rate the videos you watch; it also allows you to subscribe to your favorite video makers and upload your own videos for others to watch and comment on. Before you register your account, however, you need to think about your username.

Choose Your Username Carefully

Your username is almost important enough to deserve its own chapter. The username you choose when you register is your permanent username. There is no way to change this later. Be very careful when selecting a username. Make sure it's readable. Make sure that if someone were to mention it in a video, others would be able to type it into their browser window and find you. Make sure it isn't too similar to other popular users so you can avoid confusion.

Take this from someone who regrets his username every day. fallofautumndistro is too long, it's unpronounceable for many, and, for those who can say it, it is hard to spell correctly. When I registered my account, I never planned on using it for anything other than online video storage. I embedded the videos on my personal website (an online zine distro called Fall of Autumn, which is how I chose the name) and forgot about my account until my first video, a documentary about offset-printing techniques, was featured in the DIY and HowTo category (now called the DIY and Fashion category).

By the time I realized my username was difficult to remember, spell, and say, it was too late; I had already received a number of views and subscribers and was stuck with it. Don't repeat my mistake; keep your username short, readable, and free of any numerical characters attached to the end.

Michael Dean's username is kittyfeet69. He also says he wishes he'd picked a different name. ("kittyfeet" is from one of his websites, and "69" is the street address of his childhood home. He picked it because the username Michaeldean was already taken—by someone who has never uploaded a video! Michael thought he could pick a different username later, because on many social networking sites, you can change your username.)

If you later really want to change your username, you can simply create a new channel, in other words, a new account. YouTube allows you to create as many accounts as you'd like with the same email address. I advise against this if you have already built up your channel, though. It's a lot of hard work to build your brand, and you wouldn't want that work wasted by moving to a new, and empty, channel.

Usernames are not case-sensitive. Thus, fallofautumndistro will point to the same page as Fallfautumndistro or FallOfAutumnDistro or even FALLOFAUTUMNDIS-TRO. But it can be confusing to newcomers to put a lot of capital letters in the middle of your name, because they may think they have to type them.

One consideration is that it sometimes is prudent to use a capital letter at the start of the name, because it's your name on YouTube, and there is a psychological effect to this, as in "I take myself seriously enough to put a capital letter at the start of my name because I am a proper noun." But again, people won't have to type that letter as a capital letter; it will work typed as a capital or lowercase letter.

Sign Up for Your Account

The sign-up process is straightforward; you'll be asked to provide some minimum information to set up your account. Once your account is set up, you'll be able to provide more detailed information for your profile to display. (Much of this information is optional, but providing it makes people take you more seriously, because they will feel they know more about you. Channels with just a username, no photo, and no info almost never get many subscribers.)

So, let's get started.

First, click the Sign Up link found at the top of any YouTube page (Figure 4-1).

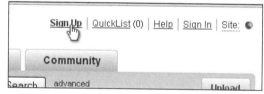

Figure 4-1. Click Sign Up to get started.

After clicking the Sign Up link, you'll be taken to the registration page (Figure 4-2).

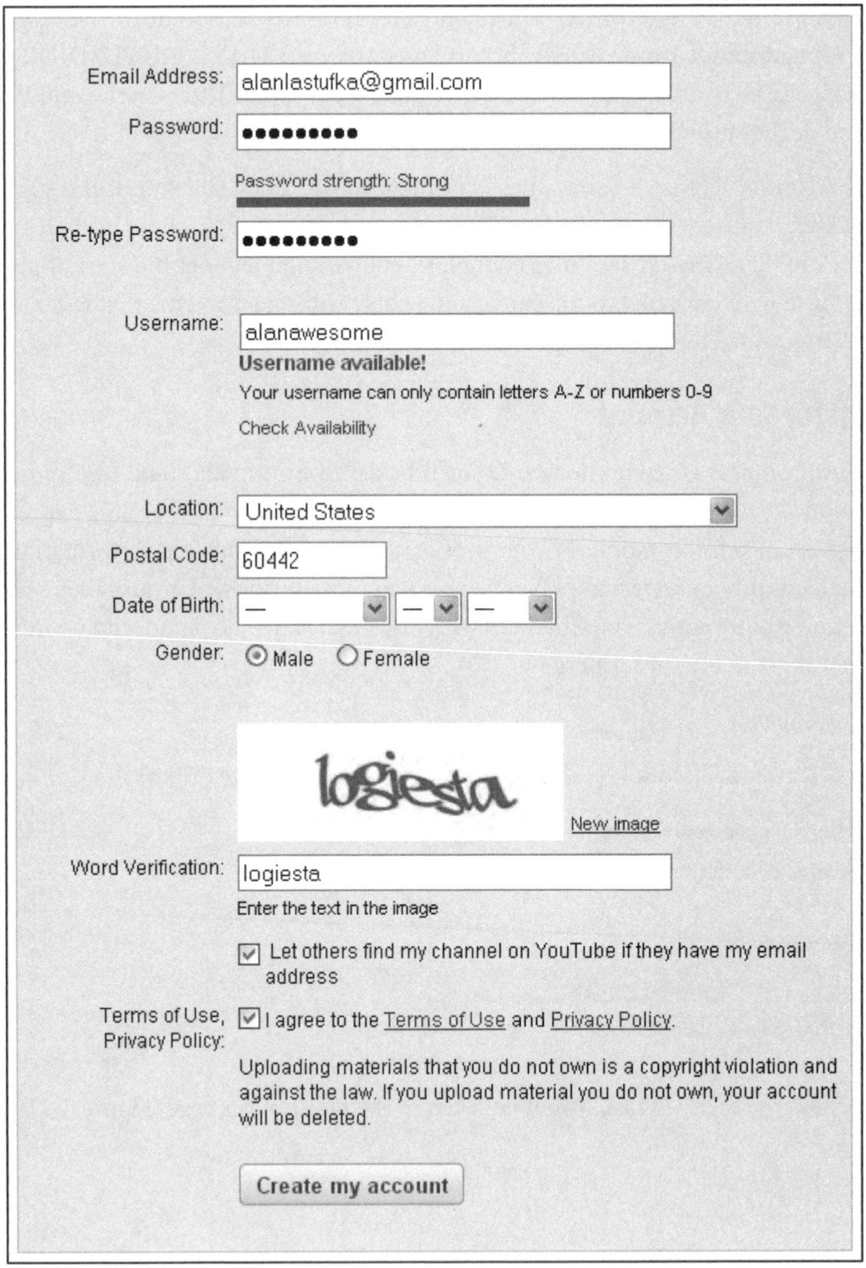

Email Address: alanlastufka@gmail.com

Password: •••••••••

Password strength: Strong

Re-type Password: •••••••••

Username: alanawesome

Username available!

Your username can only contain letters A-Z or numbers 0-9

Check Availability

Location: United States

Postal Code: 60442

Date of Birth: —   —   —

Gender: ⊙ Male   ○ Female

logiesta

New image

Word Verification: logiesta

Enter the text in the image

☑ Let others find my channel on YouTube if they have my email address

Terms of Use, Privacy Policy: ☑ I agree to the Terms of Use and Privacy Policy.

Uploading materials that you do not own is a copyright violation and against the law. If you upload material you do not own, your account will be deleted.

Create my account

Figure 4-2. Registering a new account is quick and painless.

You will first be asked for your email address. YouTube uses this email address to confirm your account, so make sure you've entered your correct email address with no typos. YouTube will also automatically send notices of new subscribers, new video comments, and other optional alerts to this email address, so make sure it's one you check often. (You can change your email address later if you want, for instance, if you get a new email address or if you start getting so popular on YouTube that you want to set up a dedicated Gmail address for YouTube-only communications to make it easier to manage all your incoming information. You can also later set up when you get email alerts if you feel overwhelmed.)

Next, you will be asked to create a password. Your password should be something difficult to guess, so don't make it your first name, your hometown, or anything that could be guessed by possible hackers. It's best to combine letters and numbers for your password. YouTube automatically rates your password's strength as you type it and requires you to type it twice to ensure that you entered it correctly.

- **Example of a horrible password**: password
- **Example of a weak password**: magic
- **Example of an OK password**: mag9ic5
- **Example of a very strong password**: 83ka0j3Ahm29n30

Then, it's time for the all-important username we discussed a moment ago. Again, take a moment before committing to a username. Think about it, because you're going to have to live with it.

Once you type in the username you've chosen, YouTube will check its availability. If it's already taken, you'll have to pick another one. If you're really set on the name, you can just add some numbers at the end of it, but check first to see whether the person who's taken it is very popular. If they have a lot of subscribers, you might want to not use a derivative of that name; you might want to pick something completely unrelated.

 (I used alanawesome in the example shown in Figure 4-2 as my username. I had to go through the account creation process and set up a new account to show you how it's done, but I'm not going to use this account for anything.)

Next, YouTube will ask for your location, postal code, and date of birth. It does this to confirm your age. YouTube's terms of use state you must be 13 years old or older to register an account. Those who are younger than 13 years old should have their parents set up and administer their accounts.

LOCATION AND PRIVACY

If you live in a very small town and are worried about people finding and bothering you (especially if you're female or already a little famous), you might want to enter a general location rather than a specific location. Michael Dean does this. He's an odd bird…very into being in the public eye but also a very private person. This is partially because he has had fans of his books and bands show up on his doorstep uninvited. On YouTube, he used "Los Angeles Area" rather than the very small town an hour or two outside Los Angeles where he actually lives.

Back when he was on MySpace, which shows your actual town based on the ZIP code you enter, he entered a ZIP code of a town a few towns over. But on YouTube, you can just enter your area. Many people will put the nearest large city that people have heard of if they live in a small town.

While many people like to put a place near where they're from to give viewers some idea of where they're from, you don't have to do so. I use "selloutville," and I've seen many people give answers like "Funky Town," "Anytown," "Smallville," "Middle Earth," or "Deep Space 9."

You also might not want to enter your real hometown, because "place of birth" is one of about four pieces of information someone needs to steal your identity. And don't say "I don't have enough money to make it worth someone's while to steal my identity," because identity thieves often steal identities of broke people and just use their information to get credit. Even if you have bad credit, criminals can steal your identity and use it to their advantage. Don't give them help doing this. And don't give potential stalkers help finding you or people close to you.

You will then be asked to provide your gender. Providing this information has no effect on your account; it is simply for demographic purposes. Word verification is next; this is a form of the CAPTCHA code, which will be discussed later in this book. In short, entering this code proves you're a human registering your account, not a computer program attempting to register numerous accounts.

You then see two checkboxes; the first is optional and allows users to find your channel by searching your email address if they know it. We're here to be seen, so I suggest making it easy for anyone to find your account. Go ahead and turn on this box if it isn't already. And finally, you must turn on the box that says you agree to YouTube's terms of use before you can create your account.

YouTube's Terms of Use

I'm not a lawyer, so I will not try to explain or summarize YouTube's terms of use in any official fashion. I will just tell you what worked for me. I strongly suggest you read YouTube's terms of use before registering your account. The terms of use provide many of the rules you have to follow while on the site. Because you turn on this box while registering an account, you can't claim ignorance should YouTube suspend you for violating one of its terms of use. I know it's a lot of fine print, but at least take a look at it to make sure you know what you can and can't do before proceeding. And please realize that these terms of use change at YouTube's discretion, without warning. Tips I may suggest in this book may not violate YouTube's terms while I'm writing this but may at some later date if YouTube adds terms. Books are set in stone (well, they are when still printed on paper), but the Internet is a moving target. Please keep this in mind. After turning on the "I agree…" box, you can now create your account.

Once your account is created, you will automatically be logged in and taken to the screen shown in Figure 4-3, which will instruct you to confirm your email address.

Figure 4-3. You must confirm your email address after registering.

Log in to your email address and find the email YouTube sent you.

> **Note** If the confirmation email doesn't arrive after a few minutes, check your spam folder, your junk folder, and your trash folder. Some overly aggressive spam blockers will send any automatically generated emails into the refuse heap—"Send them to the corner to think about what they've done." If this is the case, mark it "not spam." Let's hope your overly vigilant spam program will learn its lesson and *not* send future emails from YouTube to the trash. If you have this problem over and over, add ***@youtube.com** to your whitelist. (The asterisk is a *wildcard*. It means "allow any address at this domain to get through.")

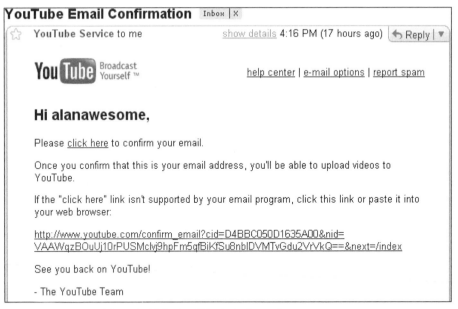

Figure 4-4. Your email from YouTube should look similar to this message.

Open the confirmation email and you'll find a clickable link to confirm your account (Figure 4-4). Clicking this link returns you to YouTube, and a "Your email has been confirmed" message appears (Figure 4-5). Now we're ready to customize your channel. The look and feel of your channel should be unique to you, so I'll show you how to access features, but I won't tell you which colors, images, or fonts to choose.

Introduce Yourself

After registering your account, the first thing you should do is edit your channel info. You'll want to provide a little basic information about yourself and about the videos you plan on uploading before anyone sees your profile. To edit your channel info, click the Account link that appears at the top of every YouTube page (Figure 4-5).

Figure 4-5. To access your account, click the Account link at the top of any page.

This will bring you to your account, which offers many options for customizing your YouTube experience. For now, we're going to focus on the Channel Info link, found under the My Channel heading (Figure 4-6). Click it.

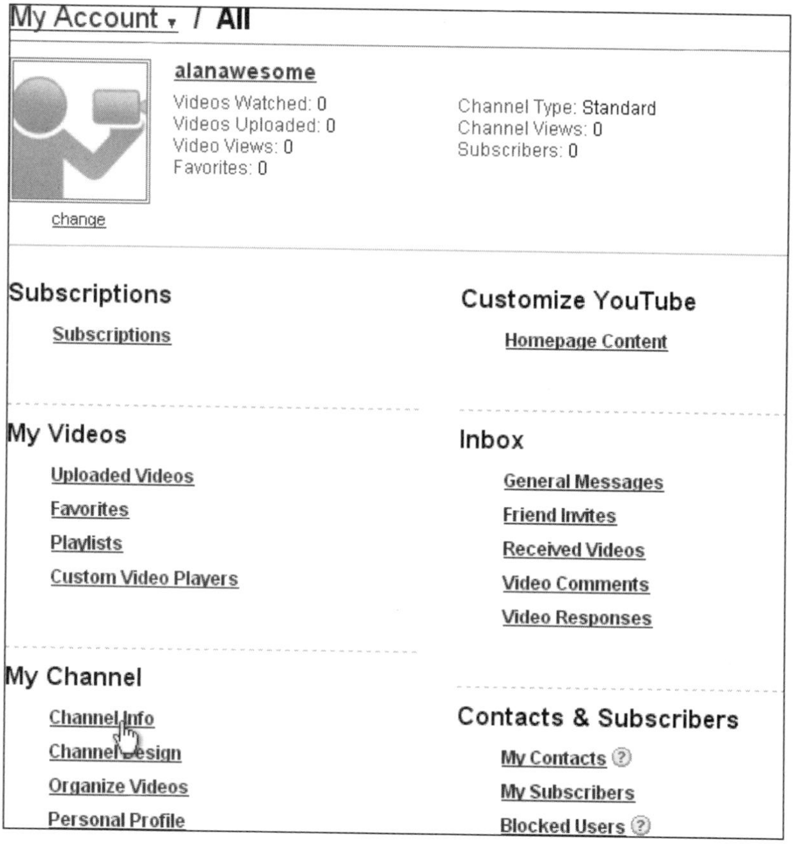

Figure 4-6. Edit your channel info by clicking the appropriate link under the My Channel options.

Now you can input some basic information about your channel.

You can't change your URL; this is set because it is tied to your username. You should memorize your URL so that you can share your channel with others via email, via social networking sites like Facebook or MySpace, or via instant messages. (We'll cover various ways to promote your channel in more detail throughout this book, but mostly in Chapters 7 and 10.)

The title refers to the text header that appears at the top of your channel. It can be your username, your real first name, or any funny phrase that represents your channel.

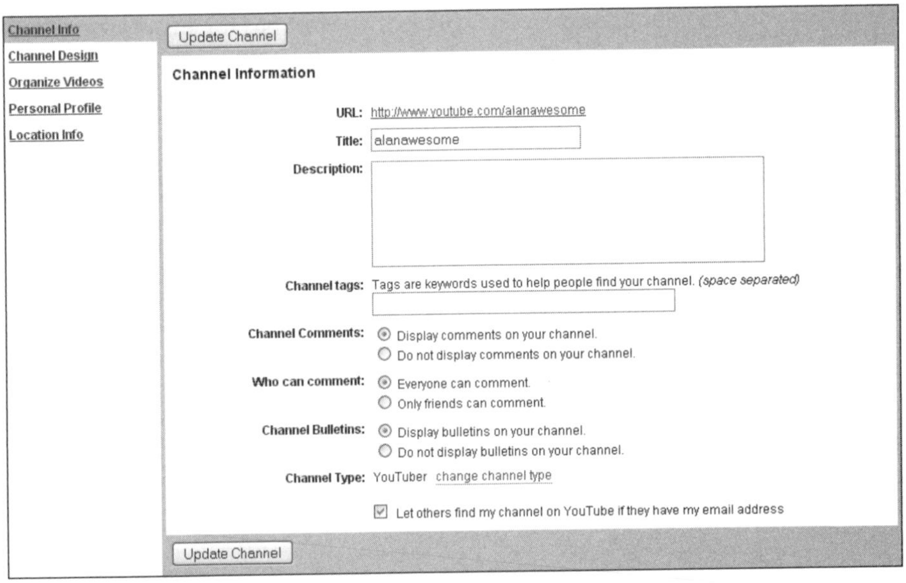

Figure 4-7. Inputting basic information about your channel is the first step after registering.

I chose "alanawesome" (Figure 4-7). You can change the title at any time, so don't feel locked in to the one you choose now.

PERSONAL INFORMATION

Keep in mind that most of the information you provide YouTube when you set up your channel and profile is public. Do not include your mailing address, phone number, or other personal information that you are not willing to share with complete strangers. If you need viewers to mail you something, pay for a post office box. They're cheap and convenient and will keep strangers from ringing your front doorbell. This may also be true of your full legal name. Between identity theft and stalkers, it's probably safer if you use a nickname or simply just your first name while online.

The description of your channel appears directly beside your channel's featured video on your profile page. The text you enter here is important for new viewers visiting your channel. Most YouTubers include a bit of personal and contact information, (email addresses or instant messenger names). If you put your email address here, rather than just letting people contact you through YouTube, expect to get a lot more spam. Your description can technically be long, but try to keep it to a readable length, and include a link to your personal website if you want viewers to read your full bio.

PRO TIP

Make your channel description personal. Make it funny if your videos are funny, or make it professional if that better reflects your videos. If you plan on uploading instructional, educational, or how-to videos, include clear contact info or websites where viewers who want to get involved can. Your channel description should be more than "Hi. My name is Alan. I make videos. Please subscribe!" That doesn't tell me anything about you or your channel, and it won't keep me around very long.

This is a better channel description:

"Hi, I'm Alan. I like punk rock and help run a book and record distribution company called Fall of Autumn Distro. I live in Chicago and just purchased my first video camera (a Sony DCR-HC48 Handycam) a year ago. I edited my first short film, *Five Stories*, nine months ago, and it went on to multiple airings on the Independent Film Channel. I plan to upload a few vlogs about my projects and also show you some works in progress for your entertainment because I'd like some constructive feedback. Feel free to drop me a line if you have any questions, and I'll do my best to share with you what I know. My email address is…."

Now, that description may or may not excite you, but at least it gives you some idea of what to expect from the channel. Your description should do the same.

The channel tags are keywords you can enter that others might search for when looking for your channel. You should include your username and any other nickname you go by. You should include words that describe your style, like "funny, sketch, comedy" for comedians or "folk, rock, acoustic, punk" for musicians.

A few other channel options will allow you to choose who can comment on your channel and whether bulletins will appear on your channel, but you will learn about those in other chapters; for now, you can leave them on their default settings, or you can experiment and change them if you're feeling adventurous.

After you personalize all the settings, click Update Channel. You can now click your URL and look at your newly updated channel.

While all the text information you entered will be there, you'll notice your channel looks really plain, simple, and, well, unattractive. Let's sexy it up a bit by pimping your profile.

Customizing Your Channel, or "Pimping Your Profile"

You may have noticed that every channel you come across on YouTube looks different. Background images, color schemes, font choices…many decisions go into making your channel look the way you want it to look. Because everyone's tastes are different, I won't tell you how to customize your channel; rather, I'll show you where your customizing options are, and I'll leave it up to you to play with them and make your channel your own. Return to your Account page by clicking the Account link at the top of any YouTube page. In the last section, you clicked Channel Info; this time, click Channel Design, which is right below Channel Info (Figure 4-8).

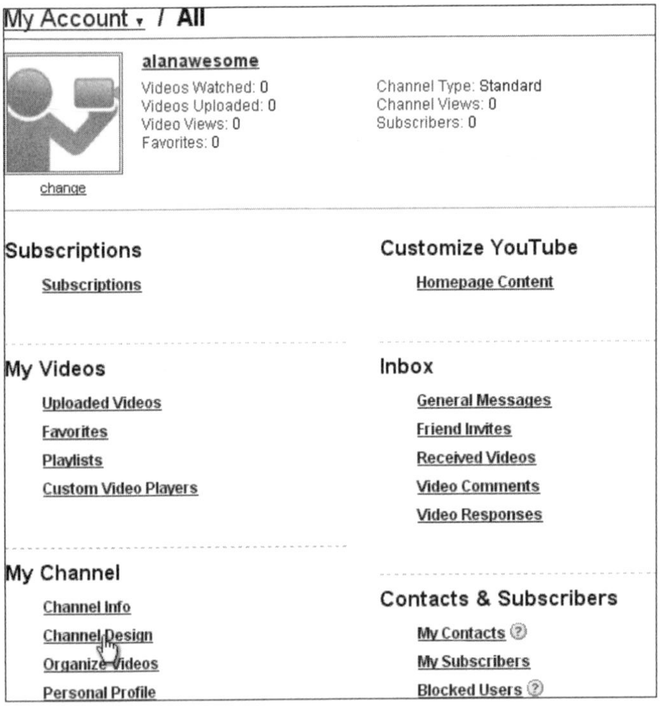

My Account ▾ / **All**	
alanawesome	
Videos Watched: 0	Channel Type: Standard
Videos Uploaded: 0	Channel Views: 0
Video Views: 0	Subscribers: 0
Favorites: 0	
change	

Subscriptions

Subscriptions

My Videos

Uploaded Videos
Favorites
Playlists
Custom Video Players

My Channel

Channel Info
Channel Design
Organize Videos
Personal Profile

Customize YouTube

Homepage Content

Inbox

General Messages
Friend Invites
Received Videos
Video Comments
Video Responses

Contacts & Subscribers

My Contacts ⑦
My Subscribers
Blocked Users ⑦

Figure 4-8. Accessing your channel design options.

You'll be taken to a new screen that offers you every option to customize the look of your channel (Figure 4-9). Adjust colors, fonts, background image, and elements of the layout. Move elements around or turn them off completely. Create a channel design you like that reflects the videos you'll upload. If your videos are edgy and dark, maybe a matching color scheme of dark grays or deep reds might be appropriate. If your videos are lighthearted, maybe light blues or oranges suit you better.

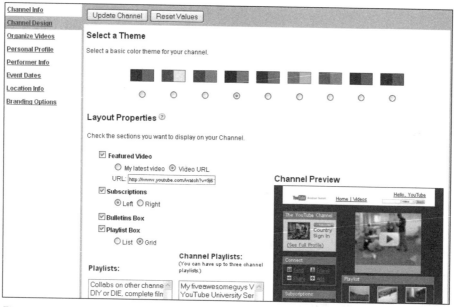

Figure 4-9. Customizing your channel's appearance.

YouTube offers a few predesigned color themes for you to choose. If you're not skilled at mixing and matching colors, choosing one of these default color themes may be right choice you. If you're really into having control over every aspect of your channel, you can pick any colors you'd like using YouTube's color picker (Figure 4-10).

Figure 4-10. Design with any color you'd like using YouTube's color picker.

As you choose your color and design options, a real-time preview will update. This will give you a pretty good idea of what your newly designed channel will look like. Again, I can show you where these tools are, but I'm going to leave the actual designing up to you. Make your channel unique, but be sure it is readable. If you choose to put orange text on a pink background, no one is going to kill their eyes to read your channel description; they'll simply click the Back button and move on. Readability on a computer monitor is important, especially if you want people paying attention to what you're saying.

Note I cannot stress this enough: You want to keep your page readable, and you don't want it to hurt people's eyes and drive them away. One problem that most people encounter when they finally get to do a little bit of "web design" is that they do things because they can, not because they should. MySpace is an excellent example of this. Many MySpace pages are so garish that you can't even read the text on them. YouTube fortunately doesn't give you as many choices, so it's hard to make an unreadable page, but try to make a page that looks good, which is not always the same as making it look really fancy.

After you design your channel and have it looking the way you want, click Update Channel, found at both the top and bottom of the Channel Design options page. Then, show it to a few friends or your brother or your mom (keep in mind, people who love you will give you an "A for effort"; strangers on the Internet are likely to be more critical). Get some feedback to make sure it's not too hard on the eyes. You'll also just naturally want to show off your new channel design.

Your channel design can be changed at any time. Change it with your mood if you'd like, or, if you have a special video coming up, change the colors to go along with your new video's theme. Stagnant channels are boring. I usually redesign mine about once a month. Doing so keeps things fresh and lets people know I'm still around and still investing time in the site, even if I haven't put up a new video that week—*especially* if I haven't put up a new video that week. But you should spend more time making videos and networking than you do pimping and re-pimping your channel.

Note YouTube Partners, users who share in ad revenue dollars, have additional channel design options and elements that will be covered, along with other information about the YouTube Partnership Program, in Chapter 11.

Subscribe to Other Channels

So, now your channel is looking foxy, you've written a great channel description, and you're ready to move on. Now what? Well, if you have a video you'd like to upload, you could skip right to Chapter 5 and start broadcasting yourself, but if you're like most new users, you don't have a clue what your first video is going to be about. It might help, then, if you watched a few videos first.

Finding videos to watch on YouTube is easy—they're all over the place!—and I covered searching for videos on specific subjects way back in Chapter 1. Most days you will log in and start off on the front page, *www.youtube.com* (URL 4.1). The YouTube editors feature 12 videos at any given time on the front page, but these are mostly hit or miss. You won't always enjoy what the editors have chosen. Instead, it's better to begin subscribing to the channels you come across that you enjoy.

When you first log in and visit the front page, your Subscriptions box will be empty (Figure 4-11). However, as you watch various videos (including those linked in this book), you'll come across numerous channels to which you want to subscribe. When you subscribe to a channel, any video made by that channel will appear in your Subscriptions box. This is a great way to make sure you don't miss any videos from your favorite users, but it's also a sure way to have something new and entertaining to watch every day when you log in. Subscribing to others is also a way to get them to subscribe to you, and having a lot of subscribers, especially ones who actually watch your videos, is a big part of working toward becoming a "star" on YouTube.

> **Subscriptions**
>
> **You haven't added any subscriptions yet.**
> Click the "Subscribe" button on any video watch page or channel page, and when your favorite channels upload new videos, they'll show up here.

Figure 4-11. Your Subscriptions box will be empty until you subscribe to a few good channels.

You can subscribe to a channel in two different ways; the first way is to visit a channel page and click the Subscribe button above the channel description.

Your second choice is to subscribe while watching any video without interrupting playback. If you enjoy what you see enough to want to subscribe, simply navigate to the right side of the video page, and click the Subscribe button there.

Figure 4-12. Subscribing to your favorite channels consists of one mouse click.

Figure 4-12 shows subcribing by visiting a channel page and Figure 4-13 shows you how to subscribe while watching a video.

Figure 4-13. You can also subscribe right from the video page, without interrupting the playback of the video.

You don't have to subscribe to the users mentioned in this book, including me (fallo-fautumndistro) or Michael (kittyfeet69), but you could do worse. The YouTubers I interview and mention in this book make consistently solid videos, interact with the

YouTube community, and occasionally produce true art. They've inspired countless videos from thousands of YouTubers by way of video responses (more on those in Chapter 8) or parody videos. They're users who routinely promote or feature others in their videos and will therefore continue to expose you to new, great channels.

You're Not Alone: Make Some Friends

In addition to, or instead of, subscribing to a channel, you can add that user as a friend. Your friends' videos don't show up in your Subscriptions box (unless you're also subscribed to them). However, being friends with another user allows you to share your videos privately with them (more on private videos in Chapter 7), and it means all the bulletins you post will appear on their channel. And vice versa.

To add someone as a friend, simply visit their profile, and navigate to just below their channel description. You will find a section titled "Connect with…" containing an Add as Friend link (Figure 4-14). After you click it, that user will be notified and will have the option to accept or reject your friend request.

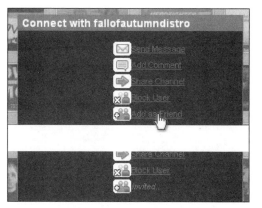

Figure 4-14. Add friends by visiting their profiles.

Other than sharing private videos or posting bulletins, the friend function on You-Tube serves one other important purpose. You can choose to allow only your friends to comment on your channel page and/or on specific videos. I advise against this if you're trying to reach the largest audience possible. But enabling this on a specific video can be a useful feature if you begin receiving a lot of spam or hate comments on that particular video. We'll cover setting this "friends-only" option in the next chapter when we discuss uploading your video.

What's Next?

As you watch others' videos and start making a few friends, you'll start itching to upload your first video. Maybe you've been inspired by a video you've just watched, or perhaps you want to vlog about your day and share some wonderful news with your new friends.

In the next chapter, we'll show you everything you need to know about uploading your video and priming it to be seen by a decent-sized audience.

5

Broadcasting Yourself: User-Generated Content

Alan Lastufka

Uploading Your Video Treasures

Never before in the history of the world has it been this easy for noncorporations (you) to reach people globally through video communication. It's pretty darn exciting. You have a worldwide broadcast tower right there in your bedroom, so let's get you started broadcasting to the world.

We've taught you how to write and direct your video. We've gone over how to edit and render your video. Then, in Chapter 4, you created your very own channel. Now it's time to have some fun, upload your video, and Broadcast Yourself...to the world.

Note "Broadcast Yourself" is a registered trademark of YouTube.

Once you've made your rendered video file (Chapter 3), you can upload it from any page on YouTube as long as you're logged in. At the top of each page is an Upload link (Figure 5-1). Clicking this link takes you to the Video Upload page. On this page, you will set all the information about your video, including the title, the description, and some keywords (Figure 5-2).

First is the title. Arguably, the title is the most important factor for viewers when choosing whether to watch your video. Don't just name your video "Yesterday" or "Alan's Birthday Party." Try to give your video a dynamic and interesting title. If you attended Alan's birthday party yesterday and your video captured the fireworks you set off at Alan's party, try naming the video "Explosions at Alan's Party!" or "I set things on FIRE!" Titles like those will pique viewers' curiosity or intrigue their sense of danger.

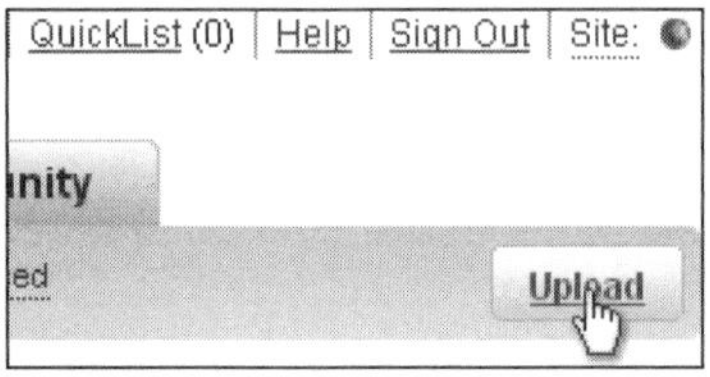

Figure 5-1. The Upload link appears at the top of every YouTube page.

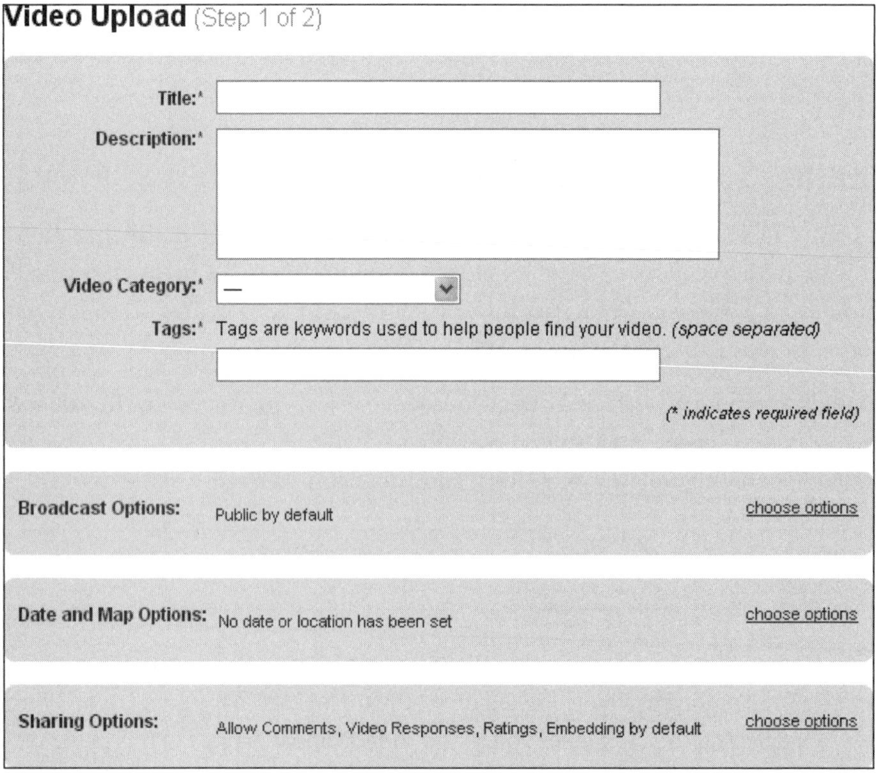

Figure 5-2. The Video Upload page offers numerous options for providing information about your video.

Don't mislead viewers with your video title, though; rather, rope them in with it. Tell the truth, but tell it in a way that makes people want to look. Deceptive titles are like deceptive keywords or thumbnail frames (covered in a minute)—people may click your video once, but after they realize they have been duped, they won't return. You will lose subscribers with deception, not gain them. And remember, a main point in having videos go viral is to gain more subscribers, especially to gain more subscribers who actually look at your videos.

Following your awesome title is your video's description. Search functions can't figure out what your video is about by watching it, because search software can't watch videos. You have to tell YouTube's software what your video shows. Choose your words carefully; a large portion of the search results on YouTube are determined by the video's description. If you simply write "Check this out!" for your video's description, the search software won't direct viewers who search for *fireworks*, *party*, or *birthday* to your video, and you'll lose out on a lot of potential views. Be honest, but be colorful.

You also need to use your video description to tell your viewers what the video is showing. Your video may be abstract, arty, or a response to a video the viewer hasn't seen. Don't leave that viewer scratching their head; provide links to any inspiration (video or text) for your video. Explain your video if it's abstract or silent or just plain experimental. (It's especially important to explain if it's silent, so people don't think "something is wrong" when they don't hear anything.)

The video description is also where you should provide web links (URLs) to any related sites. If your video is a collaborative project, include links to the other YouTubers who appear in the video or who helped you write it. (This is important. When people work for free, all they get is the credit and the link. Don't forget to name them and link to them.) If your video is about a charity, a corporation, or a news story, link to all the relevant contact information or research you did. If you're simply posting a vlog, link to your own personal website.

Don't go nuts; a small number of relevant links is better than 20 barely relevant links. Twenty barely relevant links will probably ensure that no one clicks any of them.

Finally, use the video description to reinforce your video. If you're singing an original song, you can include the lyrics and chord names in the description. (This is especially helpful if you release your songs as Creative Commons and encourage other people to record their own versions of them, which is a good way to get more publicity. You'll learn more about Creative Commons in Chapter 6.) If you have a CD available for sale, mention that in the description, along with directions on how and where to order. Provide as much information as possible so viewers don't have to ask you for it. Those viewers are unlikely to return to the video, even if you answer their question promptly. Don't make them ask; include all relevant information up front.

Next you must choose a video category. YouTube provides numerous categories, and your video should fit into at least one of them. Popular categories like Comedy, Entertainment, and People & Blogs are used often, so your video could be lost in the crowd. If you give it some thought and are more selective when choosing a category your video fits into, your video will have a better chance of being noticed. If, for instance, you vlog about spending the weekend tuning up your car, you could place that video in the People & Blogs category, or you could choose the Autos & Vehicles category, where your video would have much less competition. Competition for what, you might ask? Each category has its own Top 100 list. This list ranks the top 100 videos of the day, of the week, of the month, and of all time for that particular category. For popular categories, it is much more difficult to be placed on these lists than it is for the more niche categories. (You will learn more about category lists in Chapter 7.)

Finally, you're asked to enter a few *tags*, or keywords, about your video. Separate your tags with spaces. An example of a few tags for our video from Alan's birthday party would be *fireworks birthday party explosion wild fun time alan turned 25 years old*. When any of those words is searched, your birthday video is in the search results.

Beneath these fields are additional options for making your video public or private and for setting who can comment on your video and who cannot. You will most likely want to leave these options all set to their default status. If you do want to change any of these options or learn what their functions are, you can turn to Chapter 7, where they're discussed in more detail.

Processing, Please Wait

Once your video has been uploaded, it won't be available right away. YouTube will take some time to process your video (Figure 5-3). Processing involves converting your video into YouTube's special version of Flash video (*.flv*) format, extracting your three thumbnail options (which we'll cover in just a moment—be patient young 'Tuber), and assigning your video a unique URL so you can easily link to the video. The acres of mighty banks of rack-mount server computers located in YouTube's underground lair, probably in a secret, undisclosed location resembling the Bat Cave on *Batman*, do this all day, every day, automatically, like magic.

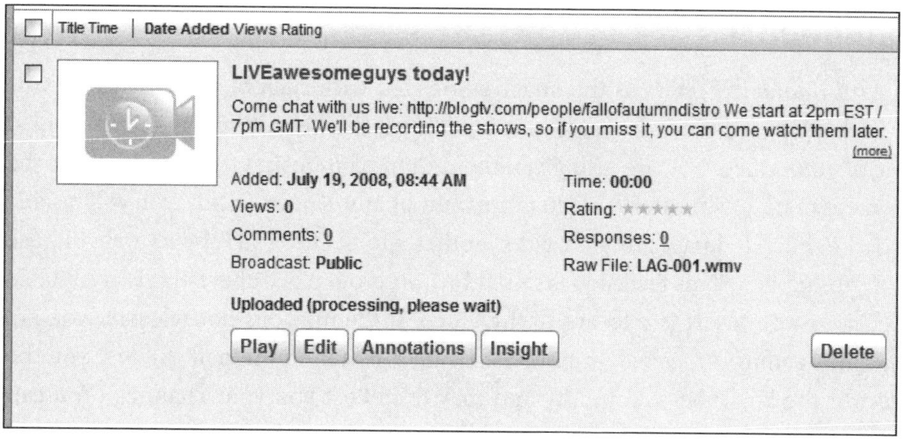

Figure 5-3. After being uploaded, your video will take some time to process.

The time it takes to process your video depends on numerous variables, including the length and format of your video, how busy the servers are processing other people's videos, and the quality of your original video. Processing times usually range from 10 to 20 minutes but may sometimes be longer. (Michael Dean has a theory that "better videos take longer," because the computers actually watch them if they're good, but this probably isn't true. That guy has a fairly active imagination.)

Once your video is processed, YouTube then refers to it as " Live!" (Figure 5-4). You can check the status of your videos at any time by clicking the Account link at the top of every page and then choosing My Videos. The status of each video will be listed underneath its title.

Figure 5-4. After processing, your video is live!

Thumbnail

The term *thumbnail* refers to the image associated with each of your uploaded videos. Each uploaded video is assigned a thumbnail, the default being the centermost frame of your video. You are also given the option of choosing the 25 percent or the 75 percent mark in your video as alternate thumbnails. Knowing this, you will be able to insert deliberate images in your video at these points for YouTube to grab and use as your thumbnail. This technique is used by numerous YouTubers and is acceptable if the image you insert is relevant to the video. If the image is not relevant, you are using a misleading, or *gamed*, thumbnail. Using a gamed thumbnail will lose you the respect of the YouTube community and may even cost you your channel. (You can learn more about cheating on YouTube in Chapter 9.)

The thumbnail you choose is an extremely important factor in whether your video will be watched. Bold thumbnails, vibrant reds, large bold fonts, or provocatively dressed young women will pull in thousands, sometimes hundreds of thousands, more views than a blurry, out-of-focus, dull thumbnail. Choose your thumbnails carefully, and keep them relevant.

To choose an alternate thumbnail for your video, click the Account link found at the top of every YouTube page, then choose My Videos, and finally select Edit. You will find your three thumbnail options on this screen (Figure 5-5).

Figure 5-5. You may select your thumbnail from three preselected options.

Favorites

As you watch more and more videos, you may come across ones you'd like to see again or show to someone else later. YouTube offers a really handy feature for this, called your *favorites*. You may add up to 500 videos to your favorites. Your favorites list is then available to you any time to review videos you liked a lot or to show your friends and family members who weren't available to see it when you came across the video. To *favorite* a video, simply move your mouse pointer below any live video, and click Favorite (Figure 5-6).

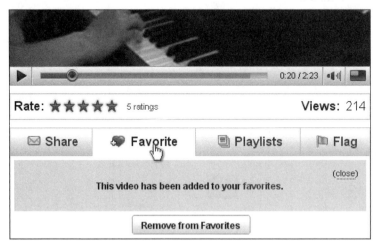

Figure 5-6. Favoriting a video to view again at a later date is easy.

A message will appear telling you the video has been added to your favorites. You can then access your list of favorite videos at any time by clicking the Account link found at the top of every YouTube page and then clicking Favorites. The videos are organized in the order in which you added them to your favorites. I really enjoy looking at my Favorites list occasionally; it acts as a sort of timeline for my hours, weeks, months, and years spent on YouTube. As I view certain videos from months ago, I recall where I was at that time and who I was actively working on projects with.

Playlists

Playlists are similar to your favorites, because they are an additional resource for saving and accessing particular videos. Playlists can even help you organize your favorites. Playlists are usually created around certain themes or genres within the videos. Perhaps you want to create a playlist for all the original music videos you find on YouTube. And say you create another for all the collaboration videos you appear in on other people's channels. These are great uses for YouTube's playlist feature.

To create a playlist, click the Account link found at the top of every YouTube page, and then click the Playlists link. To the right of the screen is a New button. Click it to create a new playlist. You'll be asked to name your new playlist. Choose a title that's short and clear so you'll remember which videos are in that playlist (Figure 5-7).

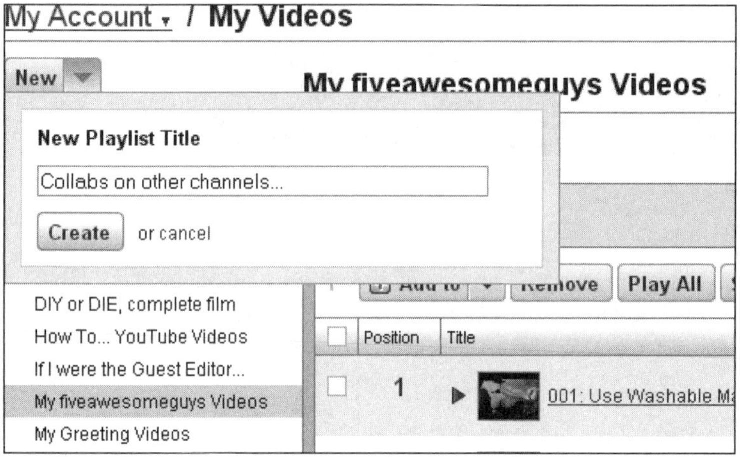

Figure 5-7. Click New to create a new playlist.

You will then be taken to a page that will allow you to edit your playlist information (Figure 5-8). Most options are self-explanatory. You'll need to briefly describe your playlist for others who stumble upon it, and YouTube will provide you with the link to your playlist so you can share it with others. You'll also need to assign your playlist a few tags, which I described earlier in the "Uploading Your Video Treasures" section.

When you're finished with all these settings, click Save Changes at the bottom of the page, and then you're ready to start adding videos to your playlist.

To add a video to any of your playlists, you will need to visit that video's watch page. Simply navigate your mouse pointer to under the playing video, and click Playlists. You will be given a drop-down menu filled with your playlist titles; select the playlist you want to save the video to, and then click OK (Figure 5-9).

Figure 5-8. Add a little information about your new playlist.

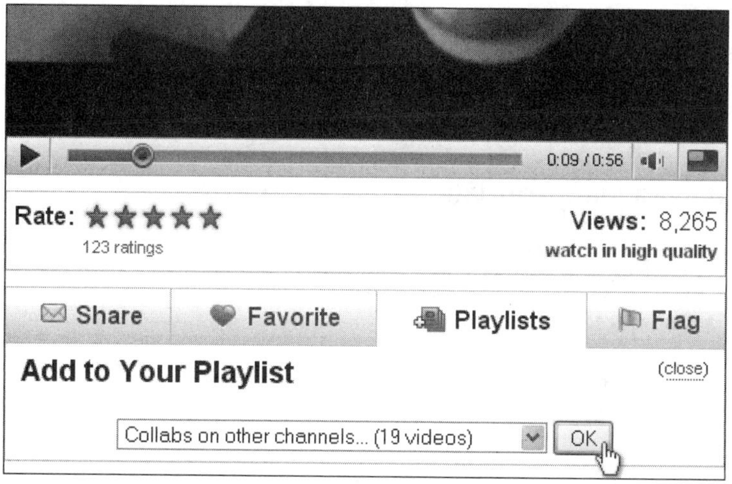

Figure 5-9. Adding videos to your new playlist is very simple.

If you have a project that is more than one video (either because the video is longer than the 10-minute limit, because you had to break it into sections, or because it's an ongoing series along a related theme), you can just send people the URL of the playlist, rather than sending them a list of many links.

I made a playlist for Michael Dean's film *D.I.Y or DIE: How to Survive as an Independent Artist*, because the video is in eight sections, which correspond to the eight chapters of the DVD. The playlist is here:

>   *www.youtube.com/view_play_list?p=66D2811AF0B5755F* (URL 5.1)

I also made a playlist of a series of comedy videos I did called *YouTube University*:

>   *www.youtube.com/view_play_list?p=BFD40BB35595AD43* (URL 5.2)

Flagging

Most of the videos on YouTube abide by the rules but occasionally you'll come across a video that you think is inappropriate. This video may show pornographic or violent images, tutorials on making dangerous devices, or unnecessary amounts of vulgarity. Maybe they plagiarize the work of someone else and call it their own. In these cases, you can help YouTube identify the offending material by flagging the video.

To flag a video, navigate your mouse pointer to below the playing video, and click Flag (Figure 5-10). You will be able to select a specific reason you are flagging the video from the new drop-down menu that appears. Reasons to flag a video might include sexual content, violent content, and so on. Select your reason from the provided list.

Figure 5-10. Viewers may choose to flag offending videos.

Once a video is flagged, YouTube says it will be reviewed by a human being, and a decision will be made either to remove the offending video or to limit the age of the viewers who can access it. There are no published rules about how this decision is made. If YouTube allows the offending video to remain on the site, it has the choice to restrict its viewers to those whose accounts state they are older than 18. If it chooses to remove the video, the original uploader will have a strike against their account. Three strikes and you're out—banned. (You'll find more about banned and suspended accounts, and what to do if it happens to you, in Chapter 9.)

The flagging system is sometimes abused when certain members team up and flag a nonoffending video repeatedly, causing it to be removed and the original uploader to be punished. (Doing this is a violation of the terms of use and can get the bullies themselves banned.) However, with the number of videos uploaded to YouTube every hour, it would be impossible for YouTube to police every video without this system in place. It's far from perfect, but most of the time it works the way it should.

If you believe your video was flagged or removed without justification, you should contact YouTube and calmly explain the situation. YouTube's contact information changes frequently. For the most up-to-date email address or contact form, visit the Contact link at the bottom of nearly every YouTube page.

Sharing

If you upload it, they will come.

Okay, in reality, nothing could be further from the truth. With so much material available to eager viewers, a lot of people rarely bother to search out new material to watch; you have to do the legwork of putting your video in front of viewers' eyes. We'll cover numerous angles for promoting your videos in the following chapters, but here I am going to go over the built-in function YouTube has provided for "pimping" your videos on third-party sites.

Found underneath each video is the Share tab. Click this to reveal your sharing options for that particular video. I recommend you share or embed your video on these sites by visiting each site. You can learn how to do that in Chapter 10, but if you're new to YouTube or any of those other sites, you may find YouTube's help invaluable.

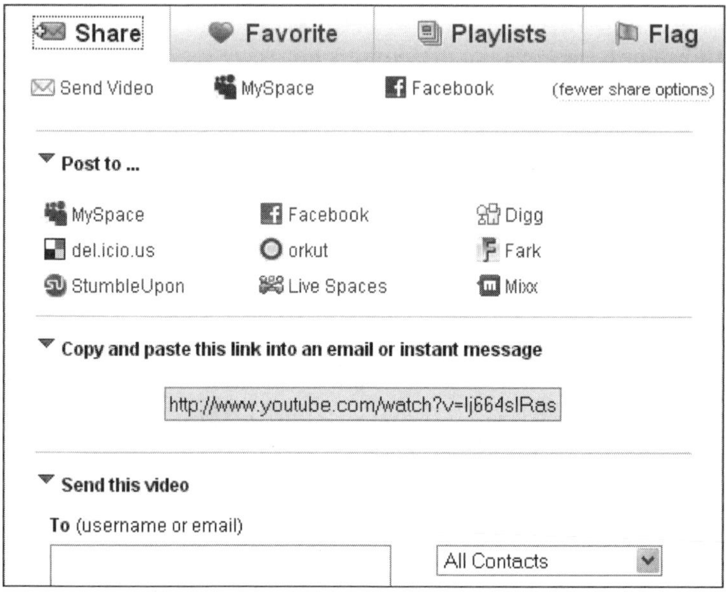

Figure 5-11. Sharing your video is easy, and it's crucial to building views.

To share your video on any of the sites listed on the Share tab (Figure 5-11), simply click the name of that site. You will automatically be redirected to the website you chose and taken to the appropriate page to post your video, assuming you are currently logged in to that site.

Choosing any of these options will embed your original YouTube video on these third-party sites. This doesn't mean views from these sites don't count on YouTube; they do. When someone watches your video embedded on MySpace or Facebook, you don't lose out on that view counting on YouTube. Embedding your video on third-party sites can only help you reach a larger audience. You can find more about building your audience in Chapter 7.

> **Note** Embedding YouTube videos into blogs run by the WordPress software can sometimes cause odd issues, such as completely reformatting your beautiful WordPress template on every page of your blog. Here's a link that explains a workaround for this issue:
>
> *http://tinyurl.com/5zgneu* (URL 5.3)

CONTACTING YOUTUBE WHILE UPSET

It's difficult to behave calmly when you're upset. After you invest a lot of time and hard work into your channel and your videos, should a video be flagged or a channel suspended, you will be very upset. Do not contact YouTube while you are upset. Being upset clouds your thinking, and the email you write while upset will be nasty and will most likely go ignored by YouTube.

When contacting anyone, it is important to remember to be respectful and calm. If something should happen to your videos or channel, YouTube does not have to fix it. It doesn't owe you anything. And YouTube editors will be more likely to help if you are courteous and respectful and clearly explain the issue you are having without cursing or placing blame.

If a video of mine were to be mistakenly flagged or removed, here is an example of what I would write to YouTube:

Howdy, YouTube,

Earlier this evening, my video titled I set things on FIRE! was removed from the site. I believe this removal was an error. Despite the title, the video does not show any criminal or negligent acts. The video did contain me lighting off legal fireworks at my birthday party. Further, the video did not include any obligatory swearing or other inappropriate content.

I'm asking you to please review the decision to remove my video. I can't see any violation of YouTube's terms of use. I even spent a good deal of time finding royalty-free music to use so the video would fall within the site's guidelines.

The original URL for the video was [insert the URL here], and the video was uploaded this morning around 9 a.m.

I would be very appreciative if you could please reverse the decision and reinstate the video. If you can't, can you possibly provide me with additional information about the reason for removal, so I can learn more and not have the same issue with future videos?

Alan

Now, please don't copy my exact wording should you have a video removed; if everyone sends in carbon copies, YouTube won't take any of them seriously. Rather, use this as a basis of some of the information you should include when addressing an issue like this. Notice I included the video's title, original URL, and time of upload. YouTube may need this information to find the video to which I'm referring. Note some things that are not in my letter: no swearing, no placing blame, and no threats.

When a video is removed from YouTube, it's never really gone; rather, YouTube simply removes it from public viewing. If you should have a video removed, calmly contact YouTube, and ask for more information. You may just be able to get it reinstated.

Permission

Here's one important last thought: Be sure the videos you upload are your own creation and contain no copyrighted materials unless you have written permission to use those materials. You have nothing to gain, and a good bit to lose, from uploading a popular music video or television show and getting a lot of views on it. The viewers aren't there to see you or your creation; you're simply providing other people's creativity at that point. You won't gain subscribers or repeat views, it will put your account in danger, and it could get you sued.

Read on as Michael covers fair use, parody, and copyright in Chapter 6.

6

Rebroadcasting: Commercial Content

Michael W. Dean

Ahhh…the joys of copyright…especially other people's copyrights. We're not lawyers, but my years of working as a semiprofessional musician and a professional filmmaker have made me a damn good "jailhouse lawyer" when it comes to intellectual property law. Here we'll talk about what you can't upload, copyright infringement, what you can upload, fair use, and parody (and dangerous common misnomers about the same). We'll explain how the rules are different for YouTube partners than for users in general, and we'll tell you where to find royalty-free music and video clips. We'll also show you how to protect your own work, with both traditional copyright and with Creative Commons.

> **Note** I'm not an attorney. Nothing in this chapter (or in this book) constitutes legal advice. If in doubt, consult an attorney. Copyright regulations vary from country to country. I can only tell you what I know about copyright regulations in the United States. Please don't email me with questions about copyright. I can't give you legal advice.

Copyright

Copyright is a right assigned to intellectual property and is indicated by the symbol shown in Figure 6-1. Music, movies, books, software, and other content are protected for a period of time from being copied and sold by people other than the creator of the content.

Figure 6-1. The copyright symbol.

Figure 6-2. U.S. Copyright Office website.

In the United States, copyrights are registered by the U.S. Copyright Office, a division of the Library of Congress, *www.copyright.gov* (URL 6.1). See Figure 6-2.

Ideas, in and of themselves, cannot be copyrighted, although an idea for a new process or machine can be patented. Obtaining a patent, however, can cost tens of thousands of dollars and take several years. Unlike patents, copyrights are not granted by the government; they are merely registered. You, as a content creator, grant a copyright to yourself whenever you create something, mark it with a copyright notice, and publish it. In this scenario, *publishing* includes putting it on the Web, printing copies, or making DVDs to be sold or given away—any method you choose to get your content out to the people. It does not have to get out to thousands of people to be considered "published," and it does not have to be sold. Usually, merely putting it on the Web for free and having it available to people, or printing copies and having them available to people, constitutes publishing.

A copyright notice is the word *copyright* or the copyright symbol, ©, followed by the year, followed by the content creator's name (or content owner's name, if it's a work for hire). For instance:

Copyright 2008 Alan Lastufka, or © 2008 O'Reilly Media

You can pay to register a copyright with the U.S. Library of Congress, but I don't usually bother doing this for small works, only for projects that involve a lot of time and money, like a book or a movie. For short works, like YouTube videos, I usually just include a copyright notice, if I even bother with copyright at all.

Almost all work owned by corporations is copyrighted, and you may run into trouble if you use materials on YouTube that is copyrighted by others, either by uploading a complete copy or even by using parts of their material as source material for your own work. This trouble can range from having the video removed from YouTube and you receiving a warning to having your account deleted to being taken to court by the owners of the material you appropriated.

A common idea is that you can do a "poor man's copyright" by mailing the media to yourself, keeping the envelope sealed, and then using the postmark to establish copyright if it ever comes up in court. This is not true and is not recognized by the U.S. Copyright Office.

Where Did Copyright Come From?

You may not know this, but, in the United States, the idea of protecting intellectual copyright is a right guaranteed in the Constitution of the United States.

This passage is from the Constitution:

> "...To promote the Progress of Science and useful Arts, by securing for limited Times to Authors and Inventors the exclusive Right to their respective Writings and Discoveries."

While this clause does not specifically use the term *copyright*, it is usually considered to lay the legal groundwork for the copyright, patent, and trademark of intellectual property, protecting the execution of "ideas" as opposed to simply protecting "hard goods." The concept of protecting the ownership of the execution of ideas was relatively new at the time; most people back then traded in goods, not ideas. But the founding fathers were big on ideas and understood that they would become more and more valuable as our small nation grew. And they understood that protecting intellectual property would encourage inventors, authors, artists, and scientists to make progress. Without the protection of their intellectual property, they couldn't make money from it and would have little incentive to get up and go to work each day.

This copyright passage is important, and it is included in the first article of the Constitution. It was considered important enough to put there. (By contrast, the rights most Americans mean when the say "my constitutional rights"—like the right to remain silent when arrested, the right to a speedy trial, the right not to be randomly searched by police without cause, and the right not to endure "cruel and unusual punishment"—were added to the Constitution four years later, as part of the Bill of Rights.)

So, as a content creator, when someone steals my content and claims that "all information should be free" (an argument that I can also see the validity of), I could easily counter with "You're violating my constitutional rights!" And people who practice unauthorized file sharing and yell that "Information wants to be free!" are often the people who yell the loudest that "My constitutional rights are being violated!" whenever something doesn't go their way elsewhere in life, often without even knowing what their constitutional rights really are. (You'll find more on constitutional rights in Chapter 15.)

Note I expound at length on the different aspects of looking at copyright and file sharing from the viewpoint of both a content creator and a content consumer in an article called "Anarchy, Integrity, and the Digital Marketplace" on the O'Reilly Digital Media site:

http://tinyurl.com/6ykjey (URL 6.2)

Fair Use and Parody

Fair use is part of U.S. copyright law that allows small portions of copyrighted material to be used without permission, like quoting a couple of paragraphs of a book in a book review. Fair use is an expansion upon the original spirit of the copyright passage in the Constitution, providing a way to help "promote the Progress of Science and useful Arts." Most works of intellectual property borrow, in some way, from previous works, at least in spirit. That is the nature of progress. Modern content creators always borrow from previous work in some way. For instance, an original blues song can be copyrighted, but the person alive today writing a blues song did not invent the blues. Someone copyrighting a poem written in iambic pentameter did not invent iambic pentameter. Someone making a mockumentary (a mock documentary, such as the movie *Spinal Tap*) did not invent the form of mockumentary.

This comes from the Copyright Act of 1976, 17 U.S.C. § 107:

Notwithstanding the provisions of sections § 106 and § 106A, the Fair Use of a copyrighted work, including such use by reproduction in copies or phonorecords or by any other means specified by that section, for purposes such as criticism, comment, news reporting, teaching (including multiple copies for classroom use), scholarship, or research, is not an infringement of copyright. In determining whether the use made of a work in any particular case is a Fair Use the factors to be considered shall include:

1. the purpose and character of the use, including whether such use is of a commercial nature or is for nonprofit educational purposes;
2. the nature of the copyrighted work;
3. the amount and substantiality of the portion used in relation to the copyrighted work as a whole; and
4. the effect of the use upon the potential market for or value of the copyrighted work.

The fact that a work is unpublished shall not itself bar a finding of Fair Use if such finding is made upon consideration of all the above factors

This is from *http://en.wikipedia.org/wiki/Fair_use* (URL 6.3):

"Fair use is a doctrine in United States copyright law that allows limited use of copyrighted material without requiring permission from the rights holders, such as use for scholarship or review. It provides for the legal, non-licensed citation or incorporation of copyrighted material in another author's work under a four-factor balancing test. It is based on free speech rights provided by the First Amendment to the United States Constitution. The term "fair use" is unique to the United States; a similar principle, fair dealing, exists in some other common law jurisdictions. Civil law jurisdictions have other limitations and exceptions to copyright."

Fair use is not black and white. A common misconception is that you can use "up to eight notes of a melody before you have to pay." This is completely untrue. Fair use has no hard and fast rule. A lot depends on context, and even then, it is open to the interpretation of the courts. Judges know the law, but how they interpret it varies on everything from how they apply the precedent of earlier cases to (since they're human) what mood they're in that particular day.

Things have changed. It was only with the advent of home computers that millions of people had the ability to make perfect digital copies of music, movies, television shows, software, eBooks, and computer games and to distribute them worldwide with relative anonymity very quickly. This scares the hell out of large corporations who trade in intellectual property (record labels, movie studios, TV networks, software companies, game companies, and book publishers). Those companies have lawyers on staff and on salary, unlike you and me. When you or I feel our rights are violated, we can say to someone "I'm going to get a lawyer and sue you." But that involves hiring a lawyer, which is expensive. If you have lawyers on staff, as the corporations do, you don't have to "go hire a lawyer." You just send an email down to the law department, and they launch into action to take care of it.

Corporations often first "take care of it" by sending a *cease-and-desist letter*. This is a letter enumerating the alleged violations and threatening swift and brutal legal action if the material is not removed immediately. This is enough to make most people take things down. Here's an article about the U.S. Air Force sending a cease-and-desist letter to YouTube: *www.boingboing.net/2008/03/08/air-force-lawyers-se.html* (URL 6.4)

Sometimes a cease-and-desist letter is not the first act of the company. They are realizing that, as more and more people see file sharing and content sharing as an inalienable right, or at least as a not-very-serious offense, cease-and-desist letters more and more often are being ignored, if not outright ridiculed. Swedish BitTorrent sharing site The Pirate Bay: *http://thepiratebay.org/* (URL 6.5)

has a highly entertaining section where they print the many cease-and-desist letters they receive from the lawyers of companies ranging from Microsoft, Apple, Warner Brothers, Sega, DreamWorks, and more. The Pirate Bay absolutely taunt and mock the letters in their replies, which are also printed on the site:

http://thepiratebay.org/legal (URL 6.6)

Keep in mind, these guys are in Sweden, where the laws differ. They're providing only torrent files, which aren't the actual media but rather files pointing to the media on other people's hard drives. The media at issue isn't just a small excerpt in a fair use capacity; it can be entire movies, whole albums, cracked working software, full eBooks, and more. And they seem willing to go to jail to prove a point.

BitTorrent is huge. Some sources estimate that between 18 percent and 35 percent of all traffic on the Internet is BitTorrent traffic, although it's hard to measure an exact amount, because of some technical issues of the protocol itself.

Things seem to work themselves out somehow in the universe. My buddy Chris Bratton (from the bands Drive Like Jehu, Chain of Strength, Inside Out, and Wool) summed it up nicely: "Welcome to 2008; movies and music are now free, but gas is $5.00 a gallon."

I'm not willing to go to jail to prove a point, at least not the point of uploading someone else's creation without permission to YouTube. Since large media corporations have lawyers on staff who are terrified (rightly so) that home computers might put them out of business, some companies skip the cease-and-desist phase and go directly to legal action. Here's an article about the Recording Industry Association of America (RIAA) suing a young girl from a poor family:

www.betanews.com/article/1063159635 (URL 6.7)

Some large corporations go after any use of any amount of their material in any context, even if it's something that a judge would probably consider fair use. They do this because they're scared, and they do this because they know that individuals do not have the means to hire lawyers for small, perceived infractions.

Note There is a lot of talk about how "copying a DVD is not the same as stealing a DVD." It's a matter of post-scarcity economics versus scarcity-based economics:

http://en.wikipedia.org/wiki/Post_scarcity (URL 6.8)

In the old economy, resources were limited. If you stole a pig from a farmer, the farmer had one less pig, so it was clearly stealing. Proponents of post-scarcity economics argue that if you make a digital copy of a movie, the maker of the movie doesn't have any fewer copies of the movie, so it may not fit the traditional definition of stealing. The moviemakers argue that if a free copy is made, that's one less physical copy they will be able to sell, so they say it is stealing. It's a complex issue, and things are still evolving. In fact, these days, the technology often moves faster than the law.

Fair Use and Remixing

Fair use gets into tricky gray areas of law when it comes to remixing. *Remixing* is basically collage work, mostly in music or video, where you take parts from several other sources and combine them in a new way into a new work. This can be done completely from other people's works or from other people's works combined with your own original content. Remixing was done before computers became common and powerful, but computers make it a lot easier to do (and easier to share), so it's become very common, and the old laws struggle to keep up.

In 1998 there was one sweeping reform of copyright law called the Digital Millennium Copyright Act (also known as the DMCA).

http://en.wikipedia.org/wiki/Dmca (URL 6.9)

This law was an attempt to bring copyright law into the computer age, but many think it unfairly favors huge corporate content owners and limits fair use by individuals. But fair use still exists, although many people claim "What I'm doing is fair use!" without really knowing what fair use is. Basically, fair use involves creating a new and useful work by combining bits of other people's original work with your own original work. Many people who do it consider it collaboration, rather than stealing, but this is open to interpretation. Many people whose work is remixed without their permission consider it stealing.

A common legitimate application of fair use is someone quoting a paragraph or two of a book in a longer printed review of that book or a scientist quoting previously published work in a new work to back up research (or to dispute the previous research). Often fair use is determined by the small amount of the borrowed material versus the larger amount of the original material and the relevant and direct expansion of the borrowed work by the new work.

Parody

Parody is making fun of people in art, often as commentary. It doesn't have to be in a mean way; it can be done by laughing "with" the subject, rather than laughing "at" them. Parody is covered, to a certain extent, within U.S. copyright law. But then again, as with most copyright issues, a lot of gray areas exist, and lawyers make billions of dollars every year haggling over these issues for people.

A common misconception is that public figures (like politicians and celebrities) are always fair game for parody. This is untrue. While they are generally more fair game than unknown people, if you say something damaging and completely untrue about public figures, you could still be sued for libel. While public figures are subject to public commentary, good and bad, statements known to be untrue and made with malice can get the commentator into trouble, involving civil and possibly criminal charges.

Parody seems to work best, and not result in litigation, when it's lighthearted rather than cruel and nasty. You can be funny and pointed without really going for the throat all the time. And it's often more effective. *The Daily Show* and *The Colbert Report* manage to stay lighthearted a lot and make their points well. (Keep in mind that *The Daily Show* and *The Colbert Report* have teams of lawyers on staff, and you do not.)

 One way that people often get away with things that are very critical of other people or corporations is by prefacing their statements with "It is my opinion that…" or "Allegedly…."

Public Domain

After many years, a copyright expires. It varies how long this takes, depending on when the media was copyrighted:

> *http://en.wikipedia.org/wiki/Public_Domain#Expiration* (URL 6.10)

Once a copyright has expired, it enters the public domain and is free to use, generally without restrictions.

Creative Commons

Creative Commons (CC) is an alternative method of licensing intellectual property that makes the work less restricted to people sharing and remixing it. Figure 6-3 shows the Creative Commons logo, and Figure 6-4 shows the website at *http://creativecommons.org* (URL 6.11).

Figure 6-3. Creative Commons symbol in the Creative Commons logo. Image taken from Wikipedia. License: "This image is ineligible for copyright and therefore in the public domain, because it consists entirely of information that is common property and contains no original authorship."

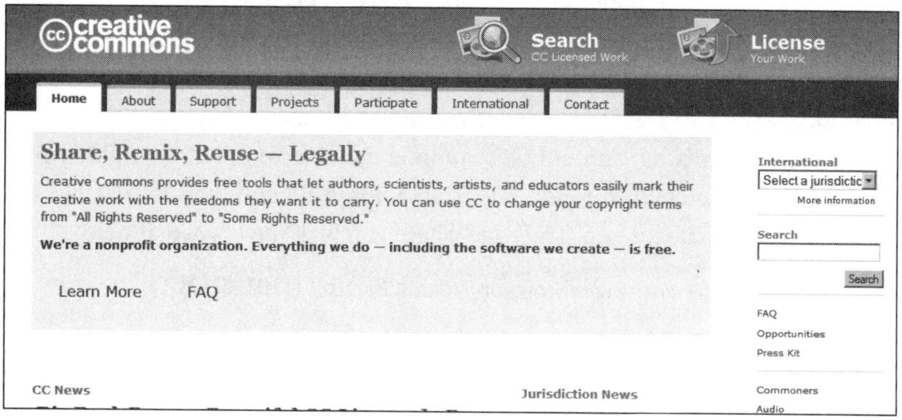

![Creative Commons website screenshot]

Figure 6-4. Creative Commons website.

Creative Commons is relatively new, formed in 2001–2002 by lawyer and Stanford Law School professor Lawrence Lessig. Contrary to common belief, Creative Commons does not exist independent of copyright but rather is a set of self-granted reduced restrictions within copyright. Using the service is free, and the organization that administers the site survives by donations. Many people are into Creative Commons with an almost religious fervor. Lessig is regarded as a rock star by many.

Here is an explanation of the many different license combinations that I find easier to follow than the explanation on the Creative Commons site.

The original set of licenses all grant the *baseline rights*. The details of each of these licenses depends on the version, and comprises a selection of four conditions:

Attribution (by): Licensees may copy, distribute, display, and perform the work and make derivative works based on it only if they give the author or licensor the credits in the manner specified by these.

Noncommercial or NonCommercial (nc): Licensees may copy, distribute, display, and perform the work and make derivative works for noncommercial purposes only.

No Derivative Works or NoDerivs (nd): Licensees may copy, distribute, display, and perform only verbatim copies of the work, not derivative works based on it.

ShareAlike (sa): Licensees may distribute derivative works only under a license identical to the license that governs the original work.

Mixing and matching these conditions produces sixteen possible combinations, of which eleven are valid Creative Commons licenses. That comes from this site:

http://en.wikipedia.org/wiki/Creative_Commons_licenses (URL 6.12)

Content creators (that could be you!) pick one of several licenses to use on their work and put a Creative Commons notice (with the name of the license being used) in place of the standard copyright notice. These range from allowing anyone to use your material in any way, for free, to the more restrictive license that lets people use the material, without remixing, and only for commercial use. You pick the rights being reserved as "one from column A and one from column B," and the site spits out the license most appropriate for you. The page where you pick the licenses is here:

http://creativecommons.org/about/license/ (URL 6.13)

Here's an example of how you'd note a particular Creative Commons license on your work, in this case, the Attribution-No Derivative Works 3.0 license:

This work is licensed under a Creative Commons Attribution-No Derivative Works 3.0 License.

Note Licenses protect you to the extent that you want to be protected. Choosing the licenses that allow the most free reuse (unlimited sharing and remixing for commercial as well as noncommercial use) probably means that your work will go further in the world and be passed around more freely and more often. The more restrictive licenses (those allowing no remixing, only sharing of complete copies, and only for noncommercial use) make it unlikely that your work will be shared as freely. If you're a stickler that your work be shared as is, without modification, or you don't want it used in a commercial fashion, choose a more restrictive license. Even the most restrictive Creative Commons license, however, is less restrictive than standard copyright-only protection.

You can also find material by other people that is licensed under Creative Commons that is free to use to remix in your own work. More on that in a minute.

 Also keep in mind that if you are a YouTube partner and making money, your videos have stopped being noncommercial and are now commercial use. Remember this especially if you are using Creative Commons content that is licensed only for noncommercial use.

How Do I Give Proper Attribution?

When remixing Creative Commons content into new content on YouTube, you probably don't need to put actual credits in the video. Most people agree that crediting and linking the content creator in the "More Info" part of the page satisfies the requirements for attribution under Creative Commons.

Why Would Anyone Even Use Creative Commons?

Why would someone bother spending a lot of time making their media and then practically give it away by allowing it to be more freely reusable than normal copyright allows?

Simple: to spread their art.

Some people just like the idea of freely sharing information, rather than owning it and charging a lot of money to use it or even refusing to let anyone else use it. Archive.org sums it up well in its motto, "Universal Access to Human Knowledge."

The world seems to be tending more and more toward that "free sharing of information" direction as computers connected to the Internet become more and more universal. A lot of people in this world are more concerned with making art than trying to sell art, and I think this is a good thing. Most people have a day job that is horribly unsatisfying. While they may make enough money to scrape by, a day job doing something you don't love crushes your soulfulness and slowly reduces you from a full human into a sort of automaton.

Doing a little art every now and then makes you a better person, a happier person, and it helps you know, just a little bit more, that life is worth living. Most people know that it is unrealistic to think they will be able to quit their day job and just do art full-time. But as more people have the tools of art and communication available to them that used to be available only to corporations and governments, more and more people want to make a little art and get it seen by at least a few people, even by someone far away who they may never get to actually meet.

Using Creative Commons means that people are not criminals for freely sharing your work. As a result, your art is more likely to spread around the world a bit and visit those places in your stead.

Copyleft

Copyleft, *http://en.wikipedia.org/wiki/Copyleft* (URL 6.14), is a movement similar to Creative Commons, but unlike Creative Commons, it has no defined legal meaning and no set licenses. It is sometimes considered an *anticopyright*, or a protest against the idea of copyright, but again, this is not clearly defined. If you want to use some media that is marked as copyleft (see Figure 6-7), make sure you read any documentation or notices included with it to see what the content creator means by "copyleft" in that situation.

Figure 6-5. The copyleft symbol.

YouTube's Stance on Copyright and Fair Use

People have gotten themselves in trouble on YouTube, both with the site and with the law, from not understanding what's allowed and what isn't. You don't need four years of law school to understand what's OK and what isn't. YouTube spells it out for you.

In Writing

In its terms of service, YouTube prohibits uploading anything you do not completely create yourself or have express written permission to use. In writing, YouTube does not allow fair use. Here:

http://help.youtube.com/support/youtube/bin/answer.py?answer=55772&topic=13656
(URL 6.16)

YouTube states the following:

What is your policy on copyright infringement?

YouTube respects the rights of copyright holders and publishers and requires all us-ers to confirm they own the copyright or have permission from the copyright holder to upload content. We comply with the Digital Millennium Copyright Act (DMCA) and promptly remove content when properly notified. Repeat infringers' videos are removed and their accounts are terminated and permanently blocked from using You-Tube. For more information about YouTube's copyright policy, read our Copyright Tips. DISCLAIMER: WE ARE NOT YOUR ATTORNEYS, AND THE INFORMATION WE PRESENT HERE IS NOT LEGAL ADVICE. WE PRESENT THIS INFORMATION FOR INFORMATIONAL PURPOSES ONLY.

A BROADCAST TOWER IN YOUR HOME

A perfect example of tools of art and communication being available to use that used to be available only to corporations and governments is podcasting, which can be done with either video or audio. Since I was a little kid, I have always wanted to have a worldwide broadcast facility in my home. Now, thanks to podcasting, RSS, and BitTorrent, I do. The weekly 1-hour episodes with me and Debra Jean yacking in our living room get 70,000 downloads a month, all over the world. And this is all without requiring a license or a transmitter and without being censored. All it took was a $200 Zoom H2 recorder, a $450 dedicated BitTorrent server computer, and an ongoing commitment to putting a few hours a week into recording, editing, and promoting.

I now have my worldwide broadcast facility in our little nest of a home. Figure 6-6 shows pretty much how I picture my house in my mind these days.

Figure 6-6. Nestenna image taken from Fickr. License: Legal remix by me of house photo by Flickr user Robert C. Wallace, www.flickr.com/photos/robwallace/ (URL 6.15), used with per-mission. This was originally marked Creative Commons noncommercial use only, but I asked the author for permission to use it in this book.

In Practice

YouTube covers itself legally in writing with the previous statement and by removing alleged violations when asked by the copyright holder. In practice, however, YouTube has a lot of copyrighted material on the site, and it often stays there until a copyright owner complains. This is not YouTube being disingenuous; it's simply that the incredible amount of content on YouTube makes it almost impossible to police it all. It's rather like the Wild West, where you'd have one sheriff and a few deputies policing an area the size of a modern county. Or larger. There just weren't enough resources to check under every bush for cattle rustlers.

Media companies' desire to extract brutal legal justice on 12-year-old girls and 78-year-old grandpas and everyone in between seems to hark back to the Wild West. In the American frontier, the few sheriffs couldn't possibly catch every cattle rustler, but when they did catch one, the trials were short, the sentences were stiff (usually death by hanging), and the hangings were very public to discourage further offenses.

YouTube deals with perceived copyright violations in several ways. If you upload copyrighted content made by someone else or use copyrighted music or video in a project you make and upload, you will probably be in violation of the YouTube terms of service.

YouTube may delete the video and send you a notice or not send you a notice. YouTube might also completely delete your account.

How YouTube deals with it may depend on how many subscribers you have and whether you're a YouTube partner. If you have five subscribers, YouTube may just delete your account. If you have 20,000 subscribers and are a partner, YouTube might be more likely to just give you a warning.

Evolving Corporate Views on Fair Use

Companies are learning that fair use is sometimes collaboration, not always stealing, and that it can actually help them "ship units" if people quote the content of those "units" without permission in a vlog or blog. But don't count on every company thinking this way.

Sometimes the legal department of a company will issue cease-and-desist orders to everybody, and if people contact someone else at that company, an exception will be made. Sometimes. Don't count on it, but it does happen. YouTube user renetto, *http://youtube.com/user/renetto* (URL 6.17) made videos of himself reading essays (in their entirety) from an NPR series of essays called "This I Believe." renetto believed that this fell under fair use. NPR sent a takedown notice to YouTube, not knowing that renetto could lose his YouTube account for this.

renetto says in this video

http://youtube.com/watch?v=M9zDyrBLDz4 (URL 6.18)

that he later spoke to an attorney, and after speaking to the attorney, renetto believed he probably had crossed a line with his readings. (Keep in mind what I said earlier that "Fair use is determined by the small amount of the borrowed material versus the larger amount of the original material and the relevant and direct expansion of the borrowed work by the new work.")

renetto contacted NPR and asked them nicely not to have YouTube take down his account, and NPR contacted YouTube on his behalf. renetto still has his account, which is good, because he has a lot of subscribers (38,259 as of this writing) and has worked hard at creating and growing his YouTube presence.

NPR even wrote him back and said, in part, "We are not at all opposed to anyone who wishes to make similar videos of their favorite 'This I Believe' essay. We simply ask that they contact us first to receive permission and proper copyright attribution from us and the individual essayist."

While this is a case of someone quickly undoing the trouble they got into from incorrectly assuming that their use of something would fall under fair use, you shouldn't assume that things will always go this much in your favor, and it seems (according to renetto's vlogs on the subject) to have caused a lot of worry.

All around, for myself, I'd say it is better to be safe than sorry and also remember that something is usually considered valid fair use if you use a small bit of someone else's material and expand on it rather than simply using all of it, even if you read it or perform it yourself.

Note YouTube takedown requests occur and are challenged between mainstream multiplatinum artists. Here's an article about Prince asking YouTube to remove a fan-shot cell phone video of Prince and his band performing a Radiohead song at a concert. Radiohead singer Thom Yorke said, "Really? He's blocked it? Well, tell him to unblock it. It's our song."

http://tinyurl.com/5g29hj (URL 6.19)

Here's a good two-step test on fair use from the Center For Social Media: *www.centerforsocialmedia.org/rock/backgrounddocs/bestpractices.pdf* (URL 6.20):

- Did the unlicensed use "transform" the material taken from the copyrighted work by using it for a different purpose than the original, or did it just repeat the work for the same intent and value as the original?

- Was the amount and nature of material taken appropriate in light of the nature of the copyrighted work and of the use?

And here is a good resource (from the same PDF) for what generally may constitute fair use:

- Employing copyrighted material as the object of social, political, or cultural critique

- Quoting copyrighted works of popular culture to illustrate an argument or point

I recommend reading that entire PDF if you want to learn a lot about fair use.

Keep in mind that as YouTube says in its own words that "WE ARE NOT YOUR AT-TORNEYS." YouTube leaves it up to *you* to make sure you stay on the legal side of the law and will do nothing to defend or protect you if someone tells them you've broken the law. This is by necessity, because YouTube would go out of business if they had to provide lawyers to give legal advice to every user on the site.

Other Ways That YouTube Deals with Copyright

YouTube also has deals with many record companies and will sometimes *monetize* (literally, to turn into money) your video with advertisements rather than take it down and give the non-YouTube share of the profits to the people who own the content you use. (When you are in the partnership program and monetize one of your own original videos, your share of the profits is 55 percent. YouTube keeps the other 45 percent. The advertisers are paying, not the viewers. You'll learn more about this in Chapter 11.)

YouTube reserves the right to do this, even if you use only a little bit of the record company or movie company's material and the bulk of the video is your original work. When a video of yours is found to be in copyright violation, the owner of the copyright has the right to monetize and take the 55 percent share, and they can do this whether you're in the partnership program or not. This is why you will occasionally see ads next to nonpartnered videos.

It's up to the record company or movie company, not YouTube, whether the video is monetized or removed. YouTube gives the companies the option; the companies review and decide.

 The companies and YouTube will probably do this only if you have a lot of subscribers. If you have few subscribers, YouTube won't bother and will probably just take down the video or even delete your account.

NO FAIR USE

Alan adds this note on no fair use for YouTube partners:

YouTube has all partners sign a contract; it is an expanded version of the site's terms of use, with its focus on monetization and copyright. While it may be OK for nonpartners to use certain elements of copyrighted material under the fair use doctrine within their own work, *it is not OK for partners to use any material they don't own without express written permission*.

Numerous partners, myself included, have even been asked, after uploading our videos, to provide YouTube with a copy of written permission from the original copyright owner. This can be as simple as an email from another YouTuber granting us permission to use a clip of their video in ours or as complicated as faxing in a signed contract with a musical artist or record company stating that I can use their music for commercial projects.

While this rule is in place, it is not always followed strictly. Partners' videos must go through an approval process before they are monetized. The YouTube staff member who OKs the video for monetization is allowed to use their best judgment regarding the video's content, rating, and use of any possibly copyrighted materials. In the past, I've seen some videos pass with more than a minute of copyrighted material included within them, while others have been pulled just for a single still image or brief clip that the YouTuber didn't own the rights to.

My recommendation is to play it safe. YouTube generally has a three-strike policy, but why risk it? Some members' accounts have disappeared after one strike or had their partnership revoked. If you must include someone else's protected work within your video, whether it's a song playing in the background or a clip of *Family Guy*, don't monetize it. It's much less likely to get you in trouble if YouTube's liability is also lowered by not capitalizing on the clip.

Getting Permission to Use Corporate Media

Generally, I don't even bother. I constantly get emails from people asking me how to get permission to use things like a clip from *Family Guy* or something from *American Idol* or a song by the White Stripes. Stuff like this is controlled by major corporations, and major corporations generally operate under the "scarcity model" and are highly unlikely to allow free use of something in a YouTube clip. Even if they were, by some freak accident of the universe, willing, you probably wouldn't be able to navigate the corporate labyrinth to find the person in charge of allowing this, and even if you were, it might take a year for them to get back to you about it.

Because of this, I find it much more sensible to use material that is allowed to be freely used, because you can find it in a few minutes online (more on this in a minute), because you're not going to get in trouble, and, mainly, because I like being a highly mobile art ninja who doesn't have to ask permission. Having to ask permission from the lumbering dinosaur who is a major media corporation is anathema to my D.I.Y. process of changing the world from my laptop. It takes too darned long.

When Someone Uses Your Video Without Permission

Sometimes, people will upload your content without your permission. People have different reasons for doing this. The most insidious is them claiming that they, not you, created the content, and this is pretty much always time to notify YouTube on your part.

CONTENT VERIFICATION PROGRAM

Alan says this: I applied for the Content Verification Program:

www.youtube.com/cvp_app (URL 6.21)

Once approved, you can remove videos from YouTube that violate your copyright with the click of a single button. It's quite powerful.

You don't have to be a partner to sign up, but you probably have to have a longer-standing account with some views on it. YouTube isn't going to trust a new account with one video on it with the power to remove others' videos at the click of a button.

If you are not approved or refuse to sign their legal documents to approve you, you can follow their general DMCA policy.

www.youtube.com/t/dmca_policy (URL 6.22)

You might want to choose your battles with this. You don't have to go after everyone just for the sake of making a stand. I own the copyright on that Steve Albini video mentioned in Chapter 2, but someone I don't know put it up here:

www.youtube.com/watch?v=5-alPPwSBRo (URL 6.23)

I didn't issue a takedown because they didn't claim it as their own. They credited the people who worked on it, so it will probably help sell a few DVDs for me. (Interestingly, the video uploaded by someone else has more views and comments than mine!) Mainly, I didn't issue a takedown order because I made the DVD it's from (*D.I.Y. or DIE*) to spread a message, rather than to make money. Here's an article I wrote about why I released it while still letting people make noncommercial copies:

www.stinkfight.com/2007/11/11/how-i-invented-creative-commons/ (URL 6.24)

In doing so, I was trying to create something like Creative Commons, before Creative Commons existed. Here's more I wrote that expands on the same subject:

www.stinkfight.com/2007/10/29/why-i-posted-all-of-diy-or-die-on-youtube/
(URL 6.25)

Finding Material You Can Use and Remix for Free

Do you like the saying "Information wants to be free"? I do. So, let's look at some places to get you some free information that you can combine with your own original information to create some new information.

Creative Commons

Creative Commons runs several sites where you can get Creative Commons content, in all media types and on all subjects. CC Audio

http://creativecommons.org/audio (URL 6.26)

is a great place to get Creative Commons–licensed background music, and samples for remixing, free. BlipTV is a video-specific Creative Commons site:

http://creativecommons.org/video (URL 6.27)

The Creative Commons–run remixing site, ccmixter.org, provides material to remix and provides a forum to upload your audio remixes; you also can use other people's remixes from that site in your work:

http://ccmixter.org/ (URL 6.28)

Flickr

Flickr has millions of images and video, some of it high-rez, much of it licensed as Creative Commons, from all over the world. My friend Skip moved to a tiny town in China and I asked him to email me photos of his town. He said, "Just search the town name on Flickr. There are already hundreds of photos of it on there." Flicker is at *www.flickr.com* (URL 6.29). See also Figure 6-7.

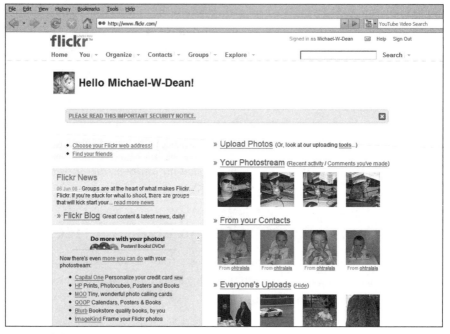

Figure 6-7. Flickr.com website.

You can find pretty much anything on Flickr. Just be sure to download the highest-rez version and read the license for that particular image. If you want to use the image for something not covered in the license, email the author and nicely ask permission. Make sure you keep the email of the content creator's consent as proof, if YouTube or anyone else ever needs to see it.

If you're a web programmer and want to start a site that works like Flickr or just want to know how it works, one of their hypersmart programmers, Cal Henderson, wrote a truly excellent geek book for O'Reilly called *Building Scalable Web Sites: Building, Scaling, and Optimizing the Next Generation of Web Applications.*

www.amazon.com/Building-Scalable-Web-Sites-applications/dp/0596102356

(URL 6.30)

Podsafe Music Network

The Podsafe Music Network is a great place to find Creative Commons music. *http://music.podshow.com/* (URL 6.31)

Wikipedia

Wikipedia lists many sources for Creative Commons audio, video, images, and text: *http://tinyurl.com/5zrxvc* (URL 6.32). Read the license on the media you intend to use to see what's allowed. You'll have to give attribution and provide a link to the people who created it.

Another great source of Creative Commons media is Wikipedia. Everything on Wikipedia—the text, the images, the audio, and the video—is covered by either Creative Commons, public domain, or is included as fair use.

Archive.org

Another great place to find material, much of it Creative Commons or in the public domain, is on this site: *www.archive.org* (URL 6.33) See Figure 6-8.

Figure 6-8. Archive.org website.

One section of the site in particular, the Prelinger Archive, *www.archive.org/details/ prelinger* (URL 6.34) is a great source of downloadable, editable versions of old industrial (training) films and educational films from the 40s, 50s, and 60s. Most are in the public domain and are free to use anywhere, for any reason, even without attribution. (The Prelinger Archive is run by a very cool guy, Rick Prelinger, so go ahead and credit Archive.org if you use stuff from there.) Use that search form at the top of the page. If you're looking for videos on space travel, search for *space*. If you're looking for videos with John F. Kennedy, search for *JFK*, and so on.

> **Note** Just because something is called "fair use" somewhere does not mean it's fair use for you to use in a different way. Nor does being called "fair use" in a particular instance by anyone other than a judge mean it is fair use. Nor does using media in a particular situation being determined by a judge as fair use mean that you can use it as your own fair use. But if something is original and the content creator labels it as Creative Commons, then it is freely useable by anyone, as long as you follow the particular Creative Commons license that it was released under.

These (mostly black and white) gems are useful in any historical piece and are particularly asking to be used in parody and comedy pieces. (One fun place to start is finding one that "speaks" to you, or seems particularly silly to you; then write and record your own narration and add that in place of the original narration. It helps if you do the work in a vocal delivery style that mirrors the "overly professional, stilted authoritative" voice that seemed to be the standard back then.)

As always, read the licensing information on the site for each particular video, because it can vary from video to video.

> **Note** Because you have the right to use a particular image or piece of video, that does not give you the right to use the likeness of the person in the image or video, particularly if you seem to be falsely endorsing something that the person would not approve of or has not been paid to endorse, especially for commercial use. This is usually less of a problem in a noncommercial use, but it can still be an issue.
>
> Also, appropriating someone's trademark in a way that incorrectly implies some sort of endorsement can likely get you into trouble, legally or with a warning, takedown, or removal of your account, even if it's in an image that was provided as Creative Commons.

Another great part of Archive.org is The Wayback Machine where you can enter any URL and see what the site looked like a few years ago. You can use it for research, or even to disprove something someone said, particularly if it was with regard to what used to be on their own website.

www.archive.org/web/web.php (URL 6.35)

The Wayback Machine is named after the time machine in *The Rocky and Bullwinkle Show*, a surreal television cartoon from when I was a kid. I loved that program so much, I have a tattoo of one the Boris Badenov characters on my chest (Figure 6-9).

http://en.wikipedia.org/wiki/Boris_Badenov (URL 6.36)

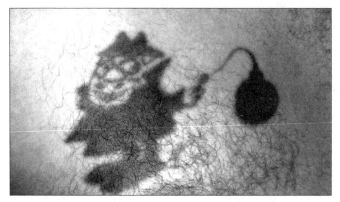

Figure 6-9. My old, very faded tattoo of Boris Badenov.

C-SPAN

The political cable channel C-SPAN is a great source of political fodder for remixing and commentary videos. According to the site,

www.c-span.org/about/copyright.asp (URL 6.37)

C-SPAN does not restrict fair use, if the following criteria are followed:

C-SPAN permits **non-commercial** use of its video coverage of **federal government-sponsored events** so long as **C-SPAN is identified during the use as the source** of the video.

Keeping a "C-SPAN" logo on the screen during the use is sufficient to identify C-SPAN as the source.

This generally unrestrictive policy regarding non-commercial use does not apply to (i) original programs created by C-SPAN, (ii) video coverage of privately sponsored events, and (iii) video coverage of other events not sponsored by the federal government.

C-SPAN **does not permit unlicensed commercial use of any** of its video programming regardless of whether the use cites C-SPAN as the source of the video. Commercial uses of C-SPAN video may be permitted under an individually negotiated license for which a license fee may be due. [See Licensing and Permissions Requests]

Nothing in this copyright policy is intended to affect any person's right to make a "Fair Use" of C-SPAN video programming.

The Bottom Line

Basically, I don't like to risk uploading content that contains work that belongs to others, at least not uploading a whole clip by someone else. If I want to use music or video in my work, I get something by a friend who they'll let me use, make music myself, or find something on one of the sites covered in this chapter that provides hassle-free media.

As for uploading complete work by someone else, you probably don't need to do this. Most things that are out there have already been uploaded by someone, and if something hasn't been uploaded, it probably will be soon.

At least one executive at a television company has been quoted off the record as saying "If it's a popular show, we'll issue a takedown. But if it's a new show that's struggling to get an audience, we'll wait on issuing a takedown, as the free publicity will help, not hurt, at that point."

But I recommend you concentrate on uploading your own brilliance, not stuff made by other people.

Going Forward

So far in the first half of the book, we've covered how to get your work written, made, uploaded, set up, and viewable. And finally now we've covered our bases with copyright (or perhaps pushed a stick into a hornet's nest, depending on your viewpoint).

It is time, young Jedi, to move it out into the world and make it sing. So, "without further Apu,"* we move on to the next chapter, "Building Your Audience."

---

* Joke appropriated from *The Simpsons* as a wry cultural comment on fair use, under what I believe to be fair use.

7

Building Your Audience

Alan Lastufka

Priming Your Channel

I am routinely asked about how to gain more subscribers. First, you can't fake it. Well, okay, you can fake it, and I'll even tell you how later in this book, but any cheat you attempt will be transparent and won't really do you much good in the long run. Second, while having thousands of subscribers is good, it's not necessary to enjoy your time on the site or even to "go viral." All it takes to go viral is for one viewer to share your video with another, who shares it with two or three others, who each pass it on, and so on. But obviously, the more subscribers you have, the greater the chance that one of them will pass your video on to be the seed that quickly becomes a forest.

Before you set out to earn more subscribers, remember that your content has to be good. I can show you how to get a lot of traffic to your profile, but if your channel description, your profile icon, or, most important, your videos don't entertain or pique visitors' interests, they won't subscribe! This is why we covered how to write, light, and frame your videos before we mentioned anything about getting your name out there.

Review some of the channel design tips from Chapter 4's "Pimping Your Profile" section. Take some time to write a compelling channel description. Make it personal, make it funny. Insert a favorite quote of yours. And make sure the featured video on your channel is one of your best.

Now you're ready to set out to actually use some of the following tips. My goal throughout this chapter will be to drive viewers to your profile page and videos. The best way to drive traffic to any web page, whether it's a blog, a new e-commerce site, or your YouTube profile, is to create inbound links. Inbound links are hyperlinked (usually blue and usually underlined) words or phrases that, when clicked, take viewers from an external site to your YouTube profile page.

The Website field in your YouTube user profile will automatically be turned into a link to your external site. You can put any number of URLs in the More Info field for a video, and all of them will automatically be turned into links. It's best to keep the number of links down to what you need so as not to overwhelm people. Keep in mind the short attention span of your target audience. I recommend just a few links to important and relevant sites, such as the websites or YouTube pages of people who helped collaborate on that video. Of course, you should also add a link to an information page with "How to get involved" on any type of activist video.

If you know how to create hyperlinks outside of YouTube, that's great; it will help in building your auxiliary web world later that you can link to on YouTube. But you are not allowed to use any HTML formatting on YouTube (this is to prevent security exploits). The good news, however, is that your username will always be hyperlinked on YouTube, so you don't have to give it a second thought. Now, let's go get your username out there....

Views and Ratings

YouTube, for the most part, is a numbers game. View counts are often used as the benchmark of a video's success or failure in this game. Of course, we don't consider a video a "failure" if it doesn't get a lot of hits; sometimes just the act of making it has its own rewards. And sometimes you'll want to make a video private when it's intended for only a few friends and family. (Read more on this in the "Privating: Covering Your Online Tracks" section.) But a lot of people do look at numbers as success or failure.

YouTube communicates how important video views are in numerous ways. First, YouTube defaults all of its top video lists to display the Most Viewed videos for each category. If you're looking for Comedy, the first group of videos displayed is not necessarily the funniest or those eliciting the largest reaction, but those that are Most Viewed.

> **Note** The Most Viewed, Most Discussed, Top Rated, and other top video lists are a great way to earn some very visible inbound links, so landing on them should be a goal. Each hour spent in one of the top 20 slots on the Most Viewed page, for instance, can earn you between 10,000 and 30,000 views!

Second, YouTube displays the view count for each video as the largest item beneath your video (see Figure 7-1), drawing attention first to how many people have watched and second to how many have commented, rated, or favorited.

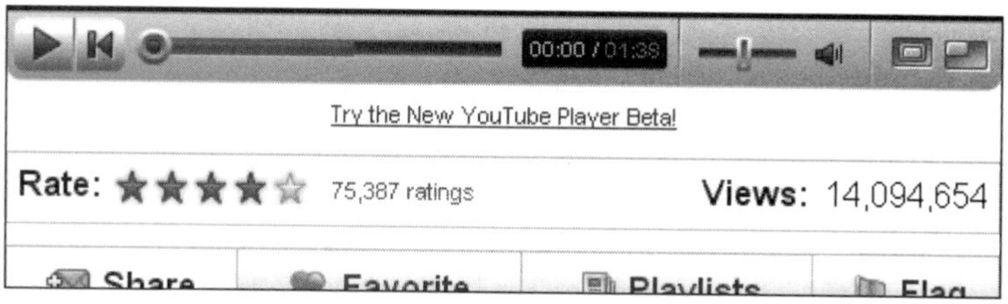

Figure 7-1. YouTube's focus is on views.

The view count has been perceived as so important that new and old users have created and shared browser plug-ins that assist in artificially inflating a video's views. These plug-ins, known as *autorefreshers*, are available for all the major Internet browsers. Autorefreshers work by reloading your video's page at preset time intervals. For example, if I set an autorefresher to reload my video page every 5 seconds, it produces 12 views per minute, or 720 views per hour. Autorefreshers are automated, which means I can let it run while I take a 3-hour nap and wake up to more than 2,000 artificial views. Those views will make my video appear to be more popular than it really is but it's against YouTube's terms of use (which could get your account suspended), and, as with most cheating tactics, it won't leave you with any sense of accomplishment.

Low numbers can be frustrating for new vloggers and video makers. It's difficult to invest hours into making a video, only to upload it and find a day later that only some 10 or 12 people have watched it. Trust me when I say this, though—we have all been there. If your content is interesting or funny and your shot isn't completely out of focus, you will gain more views over time. Faking your views will get you called out very quickly, and the majority of YouTubers will lose all respect for you.

It's best to let your views build organically. No one is an overnight success. We all have early videos with very few views, but the more you interact with the community, the more people will click through on your username and watch what you have uploaded. And because your videos rock now that you've learned proper lighting and audio in Chapters 2 and 3, people may even comment, rate, or subscribe.

Ratings are given on a five-star basis and are important because, just like with Most Viewed, there's also a Top Rated list for each video category. These lists typically favor commercial content from sponsors such as presidential candidates, Adult Swim, or the NBA. However, with a large enough audience rating our videos, we real users can place right next to the commercial content and receive the same exposure (remember that "level playing field" we mentioned in Chapter 1?).

The more ratings—and the higher those ratings are—that a video receives, the better it places on the Top Rated list and any future video search results. To rate a video, click the first, second, third, fourth, or fifth star (see Figure 7-2), depending on your opinion of the video. The ratings break down as follows:

- One star: Poor
- Two stars: Nothing special
- Three stars: Worth watching
- Four stars: Pretty cool
- Five stars: Awesome!

Figure 7-2. Rate honestly and rate often.

Comments and Replies

Commenting is the heart of interaction on YouTube. When I receive an interesting comment on one of my videos, still to this day I will click that viewer's username and watch one of their videos. I'll leave them a comment or two, and if I enjoy what I see, I will subscribe. Many other YouTubers share this habit; it's how we all continue to find new, interesting people to watch.

Viewers don't have to spam me or ask me to check out their videos or employ any other annoying gimmick; they simply interact with me by watching and commenting on one of my videos. Resist the urge to spam or ask for views. It's really a turnoff.

Most YouTubers reserve an hour or two after uploading a video to respond to the comments that come in right away on their new video. Take advantage of that information. Watch your subscriptions inbox for new videos to pop up, watch them, and comment early on the videos. If a video creator responds to your comment (something I'll stress the importance of in a minute), reply promptly. This will create a dialogue. This will get people clicking through to your profile page.

So, how do you leave a comment? There are two different types of comments: text comments on videos (found below every video on YouTube, where you comment about the content of the video you just watched) and profile or channel comments (text comments found on your profile page, if you have them turned on).

Text Comments

If you enjoy a video, hate a video, agree, disagree, or have a question, you should express that in a *text comment* after watching. Leaving comments is easy. Simply scroll down to the bottom of the page where you were just watching the video, and you'll find a text box. (See Figure 7-3.)

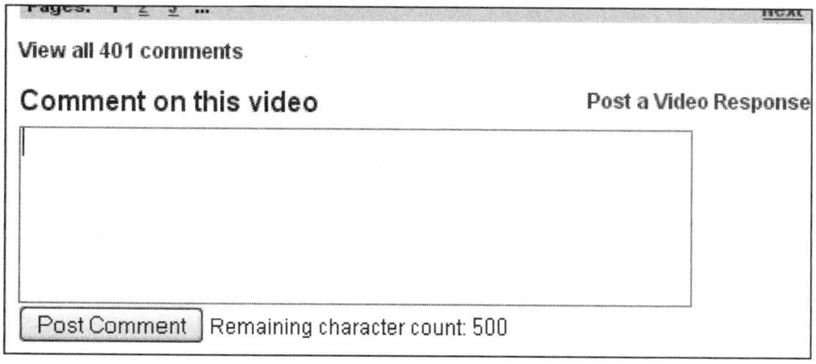

Figure 7-3. Leaving text comments is easy.

As you type your comment into the text box, you may notice the "Remaining character count" counting down how many available characters you have. YouTube limits text comments to 500 characters but lets you post multiple comments if you can't say everything you want to in fewer than 500 characters. (See Figure 7-4.)

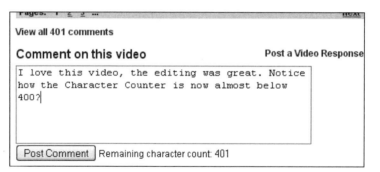

Figure 7-4. The recently added "Remaining character count" is useful.

Some users purposely break their comments up over three or four posts. This is some-times referred to as comment spamming. When users spam comment sections to ad-vertise other sites or leave cryptic "repost this three times to save your life" messages, it's irritating. However, when a friend posts multiple comments to one of your videos, they are doing so to help you. The more text comments you receive, the higher your video will place on the Most Discussed video list. Posting more than four consecutive comments on any video will lock up the comment section and keep you from posting any additional comments. YouTube then requires you to enter a CAPTCHA code to continue posting comments. (See Figure 7-5.)

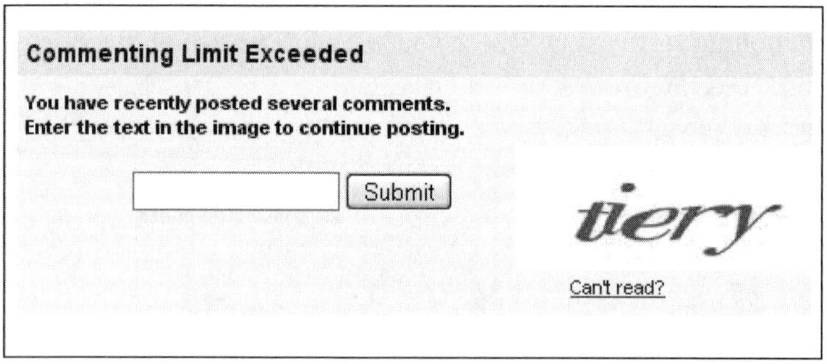

Figure 7-5. CAPTCHA codes are required after four consecutive comments.

---

* While we were going to press, YouTube added a little "Audio Preview" button below the comment box. It "reads" your comment out loud in a robotic voice. Who's that for? Blind YouTube watchers? People who can't read their own comments out loud before posting? We thought about updating all images that show the com-ment box, but decided not to. So many people dislike this feature; we're betting it will eventually be removed from YouTube.

CAPTCHA

CAPTCHA is an acronym for "Completely Automated Public Turing test to tell Computers and Humans Apart." CAPTCHA is a trademarked term, owned by Carnegie Mellon University, a nonprofit corporation. YouTube began using CAPTCHA codes to block spamming bots. The problem with spam bots, such as TellyAdder, was brought to the attention of YouTubers by popular user LisaNova, who used the spam bot herself over a two-week period. LisaNova utilized TellyAdder to leave a comment on every one of her 100,000+ subscriber's profile pages, in addition to other users. This project culminated in a video that showed her kidnapping other prominent YouTubers and forcing them to spam users in her name. A member of her production team, TheDiamondFactory, created a video shortly after the incident calling it a victory after YouTube changed its comment posting procedure to stop the bots.

YouTube never made an official statement on why it initiated the CAPTCHA code, but many believe it was because of LisaNova's and TheDiamondFactory's stunt. However, this is just speculation; YouTube could have been working on antispam measures for weeks behind the scenes, and it just happened to be released the same month as LisaNova's video.

The majority of views on your videos will be *lurkers*. Lurkers are people without accounts who watch and then move on. Lurkers don't rate, don't comment, and definitely don't make videos of their own. Lurkers are good for views, but not much else. This is why the average video views to comments ratio on YouTube is about 5 percent. Meaning, if you have 100 views, you should probably have about 5 comments; 1,000 views, 50 comments; and so on.

You want users watching your videos. You want people who will get to know, and support, you. The more invested a user feels in your channel, meaning, the more time and energy they've put in to watching and commenting and interacting with you, the more likely they are to pass your link around. Your subscribers, the regular watchers, are the ones who will rate your video every time, even if you're trying a new style of editing or writing. Your subscribers are the ones who will drop you sweet little private messages when you've been gone for more than a few days to make sure you're okay. This is where the heart of YouTube is and where you find your sense of community.

Lately, I've been having more fun interacting in my videos' comment sections than I have actually making my videos. It's imperative that you, as a video maker, reply to the text comments you receive. The more you interact with your viewers, the more you will get from the time you invest on YouTube. It's unrealistic to tell you to reply to every comment, especially once you start receiving hundreds of them, but I try to reply to at least the first 100. Viewers recognize and appreciate the dialogue.

Channel Comments

Leaving video comments on your subscribers' new videos is a great practice, not only for having fun and interacting but also for gaining those inbound links I mentioned earlier. But sometimes a user hasn't posted a new video in a while, or perhaps your comment has nothing to do with the last video they posted. What else can you do? You can leave *channel comments* (sometimes also called *profile comments*).

Channel comments are on your channel (profile) page, usually near the bottom, below your videos, favorites, etc. They're optional, however, so first I'll show you how to turn it on. Go to My Account (Figure 7-6) and click Channel Info (Figure 7-7).

Figure 7-6. A link to your account is always present at the top of each page.

Figure 7-7. Channel Info lets you edit numerous channel parameters.

Near the top you will see the option to "Display comments on your channel" or to not display them. Selecting the first button causes comments to be displayed; selecting the second causes them not to be displayed. You must select one button or the other (Figure 7-8).

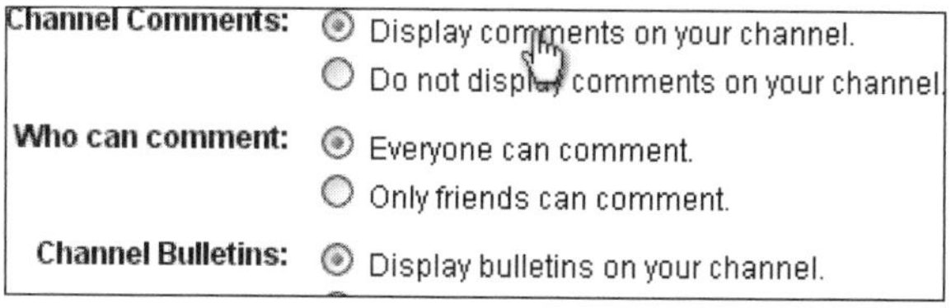

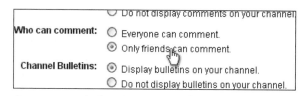

Figure 7-8. You can easily turn on your channel comments.

You can also set who can comment on your page. Perhaps you want only your friends to leave you comments; that's a great way to cut down on spam or abusive messages being left on your profile.

To set comments to friends only, first go to My Account.

Then click Channel Info.

Toward the middle you'll see the "Who can comment" setting. You can choose to let only your friends comment or to let everyone comment (Figure 7-9).

Figure 7-9. Safeguard against spam by friends-locking your channel comments.

The most common use for channel comments is to thank someone for subscribing. If you haven't opted out of the emails, you should be receiving a new email from You-Tube every time someone subscribes to your channel. This email will include a link to the user who subscribed to you. It's good form to take 30 seconds, click the username from the email, and leave them a channel comment, thanking them for subscribing.

I usually do this in batches. I'll let 10 to 20 subscriber emails pile up and then go leave comments for them all in one sitting. Just keep in mind that the same rules for "comment spam" apply to channel comments. After every four comments you leave, you will have to enter a new CAPTCHA code to continue leaving comments.

PRO TIP

Visiting the channels of those who have subscribed to you and leaving a thank you comment is yet another great way to find new channels. Those who are subscribing to you most likely share similar tastes in videos as you, because they found yours entertaining. Take a minute or two to watch what those subscribing to your channel are uploading or favoriting; you might find something new you enjoy.

More Subscribers and Friends

One additional, powerful way to gain inbound links on YouTube is to have a friend mention you in a video and link to you in the description. Occasionally, established, more popular users will upload thank you, or *shout-out*, videos. They'll want to share new channels with their subscribers. Depending on the audience of the person uploading the shout-out video, this could be a big break for your channel.

As you interact and upload your own videos, you will naturally make a few friends who enjoy your sense of humor, editing style, or lifestyle. If one of these friends happens to have a lot more subscribers than you or blows up after being featured, their shout-out video mentioning you can gain you thousands of subscribers. But don't make videos for the sake of getting someone's attention.

At the end of the day, remember to be yourself. Your audience will get comfortable with you, your sense of humor, and your video-making style. That's not to say they won't enjoy seeing you branch out and try something new, but if you get caught up in trying to make videos to please them instead of pleasing yourself, they'll know.

If you do try something different and it does well, resist the temptation to make "sequels" or to make a series. A lot of us have failed to take this advice to heart, myself included. The problem with a series, though, is that formats get boring very quickly. Look at why we have all stopped watching so much TV. Formats are predictable and unexciting. So, unless you are overflowing with the talent required for sustaining a series via great writing and great acting, your subscribers will soon stop watching. And by the time you get bored, eight sequels down the line, and move on to making more original videos, you will already have lost a number of subscribers.

UNSUBSCRIBING

Unsubscribing is hard. It's not impossible, but it requires a few extra clicks on the viewer's part. I think YouTube did this on purpose. YouTube will let you subscribe from the video-viewing page, so if you like what you see, you can subscribe with one click. However, YouTube does not let you unsubscribe from the video viewing page. If I watch a video by one of the people I'm subscribed to that bores, insults, or offends me, it is easy for me to simply click the Back button in my browser and go on to another video. To unsubscribe, I'd have to click through to the video maker's profile, click the Unsubscribe button, and then click back to my subscriptions inbox. So, don't be afraid to try something different with your videos. It's not productive to worry about whether a single video will lose you subscribers, because most people are lazy and will give you a second, third, or even fourth chance before they make those extra clicks.

Promoted and Featured Videos

Now that you're earning subscribers and making friends, what comes next? Where's your "big break"? On YouTube, your big break comes in the form of being featured.

The granddaddy of features is the YouTube front page. There are 12 slots available for featured videos on the front page, and YouTube updates this, on average, twice a day. During the 6 days you're on the front page, your video will receive anywhere from about 250,000 to more than 1 million views. The views will depend on your video's title, description, and thumbnail. An attractive or mysterious thumbnail, along with a solid title, will gain you more click-throughs than a blurry or muddy thumbnail image and uninteresting title.

I was featured on the front page of YouTube in July 2007. It comes with no warning, no notice. That morning I woke up, logged in, and could barely catch my breath when my latest video's thumbnail popped up at the top of the front page. YouTube featured an animated music video I had made a few weeks prior for my friend's independent band, imadethismistake. Here's the video:

www.youtube.com/watch?v=g8yQ0T7fNtw (URL 7.1)

That video went on to almost 800,000 views and gave me my first big jump in subscribers, but most users are never featured on the front page. It's all up to the mysterious editors and their mysterious selection process.

YouTube also features videos in each of the video categories. These category features, or *Spotlights* as YouTube now refers to them, don't generate as many views as a front page feature, but they are rewarding nonetheless. Most of the active users on the site still check the video categories a few times a day and will see your video. The categories include Music, Comedy, People & Blogs, Entertainment, News & Politics, and more. Michael's *D.I.Y or DIE* was featured by the music editor and received more than 31,000 views. Here's the video:

www.youtube.com/watch?v=PtX09q9SCXw&feature=related (URL 7.2)

Category features are selected by the editor for each category. Each category has at least one editor. Some of these editors make themselves known to the community, while others do not. The editors watch and comment on videos. Editors also welcome suggestions for videos to feature, which can be emailed or private messaged to their YouTube profiles. The editors are more likely to look at something from someone who doesn't send a lot of suggestions, especially if that person has sent things the editors have featured before.

Most of the videos featured on the front page were featured in the video categories first.

Unfortunately, there's not much you can do to make sure you "get featured." Editors don't fall for compliments or *email bombing campaigns*.

Note Email bombing campaigns consist of users asking their subscribers and viewers to email their video's link to a specific editor in an attempt to overwhelm or impress that editor. Most users do this with the hope that the editor will feature their video after being linked to it numerous times. This tactic does not work.

Privating: Covering Your Online Tracks

Now that you're making better and better videos and gaining this new, and much larger, audience, you may want to "clean up" some of your weaker back catalog. You can hide individual videos using YouTube's broadcast options.

There are only two broadcast options available for your videos: Public and Private. The majority of videos on YouTube are public; they are viewable by anyone, even if that person doesn't have a YouTube account. A few users, however, create private videos or choose to *private* (hide) past videos they've made. Most private videos are usually targeted to a handful of friends or a few family members—they are not created for the general public, and you also want to spare Grandma from seeing the occasional "you suck" posted by some random idiot.

For example, you may want to share a video of your son's first word with his grandparents. This isn't exactly the kind of video most people would broadcast to the world. Or, you know that one video where you sang "The Glory of Love" horribly out of key for your girl on Valentine's Day? Yeah, that's one of my favorite reasons YouTube has a Private setting.

When you make a video private, you can select up to 25 other YouTube accounts to share that video with. The only people who can view your private videos are you and these other accounts you select. To private a video, follow these steps:

1. Go to the My Videos section of your account (Figure 7-10).

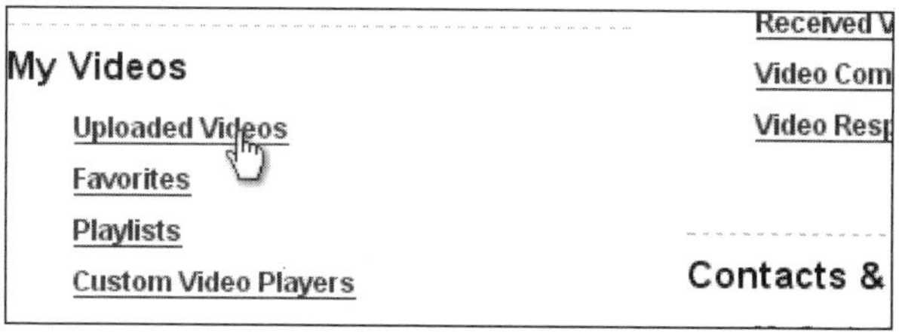

Figure 7-10. Privating videos takes only a few clicks.

2. Find the video you want to private, and then click the Edit Video Info button (Figure 7-11).

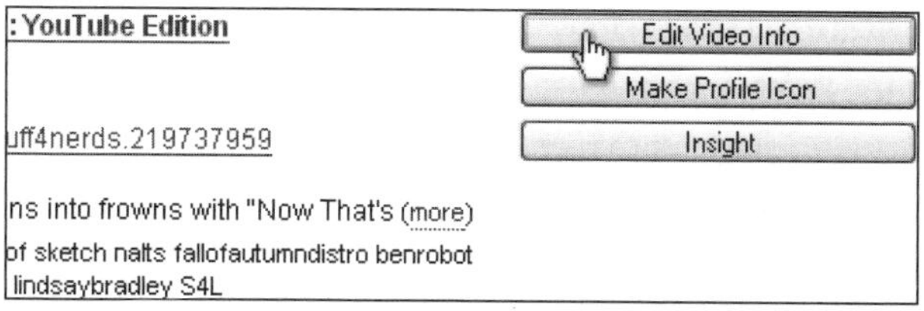

Figure 7-11. Privating is just one of the many options you can set when you edit your video info.

3. Finally, under Broadcast Options, set the video to Private. This reveals an additional section that lets you choose up to 25 other YouTube accounts, from your Friends and Contacts lists, who will be able to view the video (Figure 7-12).

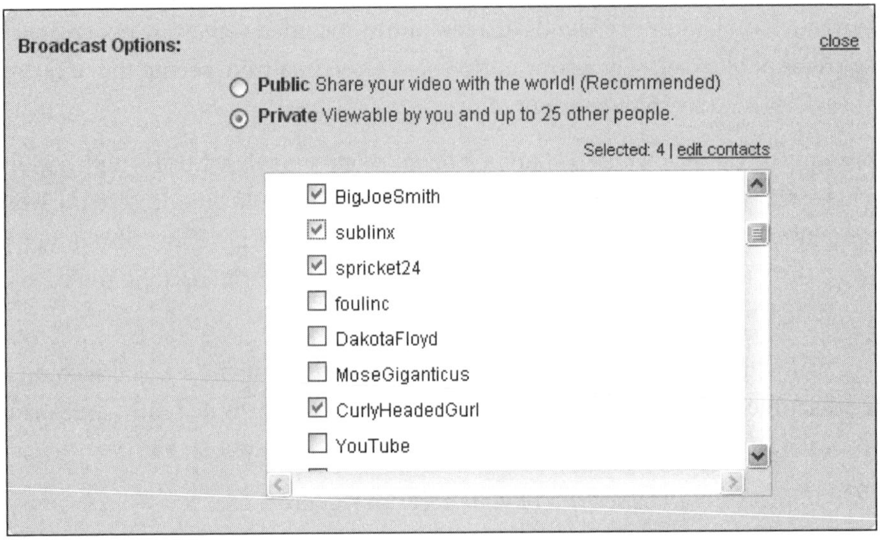

Figure 7-12. You can choose up to 25 other YouTubers who may view your private video.

4. Click Update Video Options, and you're done.

Aside from personal videos for friends and family, there are other reasons you may want to private a video. Perhaps, in the heat of the moment, you made a rant video that you think no longer reflects your opinion on a certain subject. Or maybe, as your dozens of subscribers grow to thousands of subscribers, you decide you no longer like to share some of those intimate introduction videos you posted. Or perhaps you're looking for a new job and don't think the video of you, two girls, and a beer bong will impress your future employer.

Keep in mind, privating a video does not hide it from YouTube staff. A private video is still subject to the same terms of use as public videos. This means you can't use the Private video setting to share copyrighted, violent, or pornographic videos with your friends. Removals of private videos by YouTube count as strikes against your account, the same as removals of public videos.

Playing the Game

It helps, in real life and on the Internet, if you know how to play the game. You may have also heard it referred to as *selling out*, or *buying in*—but whatever you call it, it works. I do it, and I haven't compromised any of my ethics in the process.

Everyone has something to offer others. Maybe you're good at writing short, catchy theme songs or music. MysteryGuitarMan started making some big friends on You-Tube after recording brief intro songs for popular users to insert at the beginning of their videos. These users were grateful, and a few made shout-out videos (as discussed earlier in this chapter), which drove traffic to MysteryGuitarMan's profile:

www.youtube.com/user/MysteryGuitarMan (URL 7.3)

Don't let this example of MysteryGuitarMan selling his songs to drive traffic to his profile undercut his talent and video-making skills. As I've stated numerous times throughout this chapter, I can drive traffic to your page, but you have to supply the content. MysteryGuitarMan had the content and then used his catchy, out-of-key themes to drive the traffic.

I met my first few "YouTube celebrities" after creating free banner images for them. I have a knack for graphic design, saw the need (the first few banners were horribly pixilated and designed; early partners had no idea what they were doing), and the gesture netted me a few backlinks and shout-outs in videos.

I've spoken a lot about the community in this chapter, but there is more. Collaboration videos, video responses, Advocates, and more await new YouTubers. We'll introduce you to a few of these key community elements in the next chapter.

ART KARMA

Michael adds:

There are innumerable examples in my life outside of YouTube where "art karma" has paid off for me. By "art karma," I mean doing free favors for people, without expecting anything in return, and getting a lot in return. The reason I mention things outside of YouTube is I encourage people to not get caught up in a "YouTube is all of life" mind frame. YouTube is a great place to be, and you'll need to spend a lot of time there to get seen, but you shouldn't sacrifice having a life outside the 'Tube. For one thing, spending too much time in any Internet environment is not healthy for your mind, spirit, or body. But also, you need to maintain a full life outside the Internet if you're going to have useful things to talk about on the Internet. Otherwise, you just end up in a little closed loop, and your world gets smaller and smaller.

A perfect example of art karma is how I met Alan. A couple years ago, he contacted me out of the blue and asked to interview me for an article he was doing on independent filmmakers for *Verbicide* magazine. He was interviewing four filmmakers, including me. He told me the word count he needed, and I gave him about five times the word count he needed to give him more to choose from. I also went through my answers four times to remove all typos before sending them to him. I made them perfect. I also did all of this in three days, even though I was very busy at the time, because I liked him from the start and thought it was a valuable project. I worked as hard on it as I would have for a paid project, even though it did not pay.

Alan loved my answers, and we started chatting on the phone. This book came out of those conversations, and we would never have hooked up to collaborate if I'd said "I'm too busy to do an interview right now" or if I'd done it half-assed instead of making it great. The final, uncut interview as I turned it in to Alan is here:

www.kittyfeet.com/verbicide.htm (URL 7.4)

Another case of art karma is my relationship with Chris Caulder. Chris is a fan of my *$30 Film School* book and wrote me a fan email. I treated him like an equal, rather than a fan, because he is. We chatted, and he ended up writing a free, credited sidebar for my *$30 Music School* book and did a great job on it. I helped get him hired as the paid tech editor on *$30 Writing School* and also got him a paid gig writing a book, *Digital Music–DIY Now!*, with me:

www.diynow.org (URL 7.5)

I suggested that Chris be hired as the paid tech editor on the book you are holding in your hands right now, and the company agreed and hired him.

PRO TIP

TubeMogul, *www.tubemogel.com*, (URL 7.6) is one of the most useful companion websites available to YouTubers. TubeMogul lets you track various pieces of data (such as views, ratings, comments, and more), upload your videos to multiple sites with one click, and better understand what your audience is interested in.

TubeMogul is easy; you simply sign up for a free account, enter the link to your You-Tube profile (and any other video-sharing sites you may be using), and then wait for the data to come pouring in. I find TubeMogul most useful for tracking video views from day to day and month to month. For instance, if you wanted to see the effects of all your interacting, keep track of what you do for one day and the total number of views you received that day. Then, the next day, do nothing. Take a look at the difference—you'll be surprised. I receive, on average, a 150 percent increase in views just by spending an hour a day leaving profile comments.

TubeMogul doesn't start tracking your data until you register for an account there, so head over there promptly after registering your YouTube account. You want to have as accurate, and as much, data as possible. TubeMogul will even let you compare your views (or other data) against various other YouTubers and export any data available to graphic charts (as shown in Figure 7-13) or Excel spreadsheets.

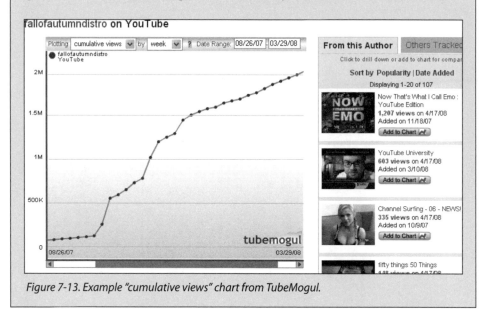

Figure 7-13. Example "cumulative views" chart from TubeMogul.

8

The Community: Where Do You Fit In?

Alan Lastufka

Networking

It sounds dirty, doesn't it? It may sound sordid and calculating and, well, corporate, but we all do it every day. *Networking* has come to mean so much more than flashy business cards and stiff suits, high-powered lunches, and rubbing elbows with folks you think might be useful to your career one day. Networking on a site like YouTube helps you make friends, some of whom may become just as important to you, if not more so, than the friends you have in "real life." As I mentioned briefly in the previous chapter, interacting and having fun on YouTube is what will keep you coming back. Anyone registering an account with dollar signs in their eyes will surely leave the site for some other pursuit within the first month. I was on YouTube for more than a year before I made a single dollar, but I didn't care. I was making art, I was having fun, and I was making friends.

Video Responses

The video response feature lets you *reply*, or talk back, to a video. Someone may post a video that you want to strongly support or that you strongly oppose. Someone may post a video that asks viewers a question or wants your opinion on a certain topic. You may just want to parody a video. These are all great opportunities to post video responses.

To post a video response, click the link Post a Video Response, located just below every video.

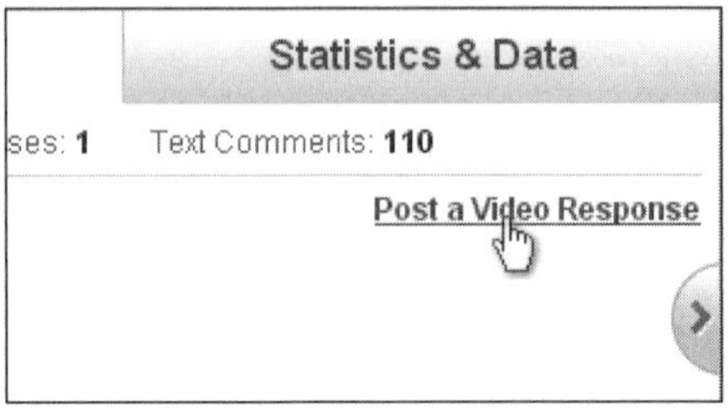

Figure 8-1. Posting video responses is fun and easy.

You can post video responses (Figure 8-1) in one of three ways:

- Record a video: If you have a webcam or your FireWire miniDV cam connected to your computer, you can record a video directly to YouTube's servers with this function. The video will be recorded in a low resolution (to conserve bandwidth and server space), but the quality will be acceptable for a simple video response. This is the most immediate and very likely the most honest way to respond to a video—in the moment.

- Choose a video: If you have previously uploaded a relevant video, you can simply choose it from a drop-down list (Figure 8-2). This is also convenient for recycling relevant video responses, something I will talk about later in this chapter.

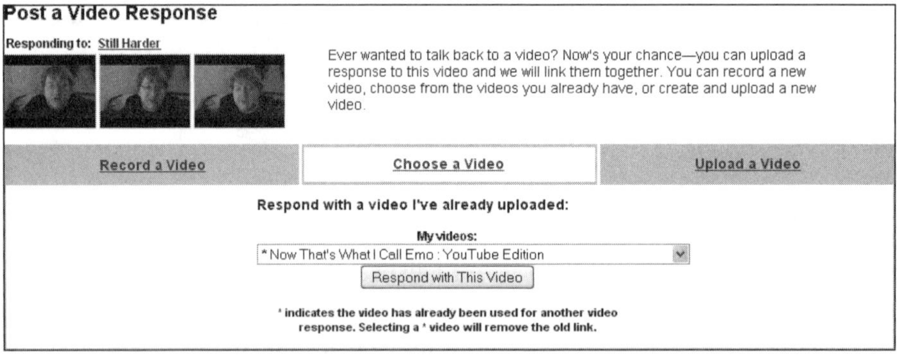

Figure 8-2. Select one of your existing videos to respond with.

- Upload a video: If you've recorded a video response but haven't uploaded it yet, you would choose the third option, Upload a Video (Figure 8-3). From here, you upload your video the same as you would upload any other video you've created. (This is probably a better option than directly capturing your video if you prefer better video resolution and the ability to edit over the immediacy of quickly "just getting your answer out there.")

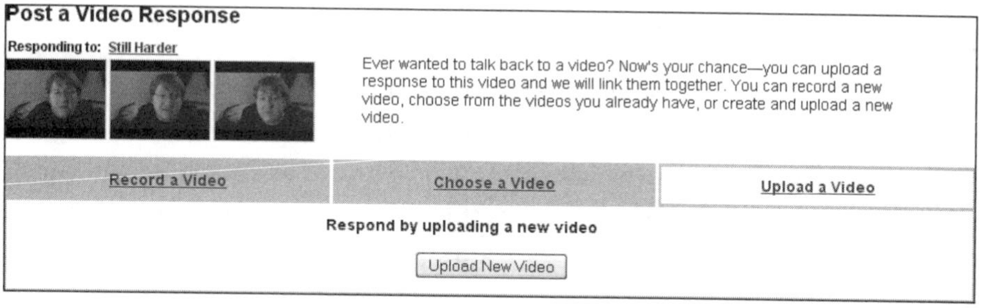

Figure 8-3. Upload a new video to leave as a response.

The majority of the early vloggers who found fame on YouTube did so by responding to other videos. Because your video response's thumbnail appears directly below the video you're responding to, as viewers watch the original video, they are likely to see, and click, your video. This is a great way to get involved and interact with the community. Video responses are a step up from text comments. You can express full thoughts without being restricted to a 500-character cutoff, your tone of voice and inflection show through, and more viewers are willing to watch video responses than they are to read someone's entire comment section.

Recycling Relevant Video Responses

Videos get few opportunities at being fresh again. Most videos reach their peak of views and comments within 48 hours of being uploaded. However, by reposting relevant videos as responses to new popular videos, your older videos get a second chance at views and discussion. I call this *recycling*.

For instance, if three months ago you posted a video retelling your own favorite childhood memory in response to someone's best childhood memory, that same video would be appropriate to repost now on a new video discussing favorite memories or memories of grade school.

By utilizing the Choose a Video option, you can leave any of your past videos as a response to any other video on YouTube. Like any other function of YouTube, this function can be abused. For instance, when a new video is featured on the front page, hundreds of people looking for views post their videos as responses, hoping for some of the spillover views from the featured video. I don't recommend doing this unless your video really is relevant to the newly featured video. This practice will only annoy the original poster and those who are leaving relevant responses if you end up leaving numerous unrelated responses looking to gain a few extra views.

Approving Video Responses

To help combat the behavior of people spamming your video response section with unrelated videos, YouTube lets you turn off automatic video responses. Selecting the Manual option lets you approve which videos will be posted as a response and which will be rejected. You can change this setting at any time for any video by editing the video info. Under Sharing Options (Figure 8-4), choose "Yes, allow video responses after I approve them."

Sharing Options: close

Comments: ◉ Allow comments to be added automatically.
                            ○ Yes, allow comments after I approve them.
                               Friends can add automatically.
                            ○ Yes, allow comments after I approve them.
                            ○ No, don't allow comments.

Comment Voting: ◉ Yes, allow users to vote on comments.
                            ○ No, do not allow users to vote on comments.

Video Responses: ○ Yes, allow video responses to be added
                               automatically.
                            ◉ Yes, allow video responses after I approve them.
                            ○ No, don't allow video responses.

Figure 8-4. Manually accepting video responses is a great way to cut down on spam.

Video Response Memes

Similar to any MySpace, LiveJournal, or Blogger meme that we've all, at one time or another, participated in, YouTube has its own unique *tag* game: the video response meme. Some memes come in the form of a survey, with a standard set of questions for you to answer and then pass on; others involve you stating "50 things" about yourself or involve "nude vlogging" or any other random qualification. Each meme video ends with you tagging a number of other YouTubers who then have to continue the cycle.

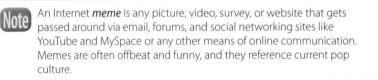

Note An Internet *meme* is any picture, video, survey, or website that gets passed around via email, forums, and social networking sites like YouTube and MySpace or any other means of online communication. Memes are often offbeat and funny, and they reference current pop culture.

The most popular video response meme to date has been the Vlog Tag Game, *www.youtube.com/watch?v=zzIMbjA5m8s*, (URL 8.1), started by YouTuber nerimon (Figure 8-5). The Vlog Tag Game requires each user to state five facts about themselves that their viewers may not already know and then tag five other users to do the same, asking those five users to leave their videos as responses. nerimon wanted to see how long each chain he began would continue before being broken by someone who received no video responses. The game grew at such a rate that YouTube featured nerimon's initial video on the front page and encouraged all YouTubers to play along.

Figure 8-5. nerimon introduces YouTube to the Vlog Tag Game.

This game has many other incarnations, and new ones pop up every day. They can be a lot of fun to participate in, and even more fun to start. If you find an older video of a meme you'd like to answer, don't wait to be tagged; make a video response, and start the game over again, tagging five of your own viewers to continue it.

Collaboration Videos

If leaving text comments on others' videos is the best way to get your name out there, then making a collaboration video with another YouTuber is the best way to get your face out there, the best way to get involved with the community, and some of the most fun you'll have making a video, I promise. Collaboration videos are any videos on YouTube that feature more than one YouTuber. Sometimes these videos tell a story, with various YouTubers playing different characters, like my Halloween 2007 collaboration with WhatTheBuckShow, SpeedyConKiwi, mikeskehan, and Hank from the vlogbrothers: *www.youtube.com/watch?v=4MVnBxx4GYU* (URL 8.2); other times collaboration videos are used to bring awareness to a cause or to unify many voices (like my tongue-in-cheek collab about collabs featuring more than a dozen other YouTubers: (*www.youtube.com/watch?v=717oxEOfERQ*) (URL 8.3).

Collaboration videos let you interact with some of your favorite Tubers while exposing your viewers to new users. When I first started making collaboration videos, I aimed to work with higher-subscribed-to channels than my own, I wanted people with more subscribers and more friends, and I had a lot of fun making collabs as a way to expose myself to these larger channels' viewers. Now that I've found an audience of my own, my collabs feature the friends I've made on YouTube and newer, up-and-coming channels that I want to help support.

Your collaboration video can be about any topic you'd like and it can tell any story you want to tell, but you should keep in mind the following guidelines when making a collaboration video, especially your first few times:

- Always have a solid idea for a video before contacting anyone about collaborating. YouTubers spend enough time as it is writing and creating their own videos; they will be much more likely to work on a collab with you if you have a plot/story/concept that you can easily explain to them. A script is nice too, if that works better. Anything is better than a vague, ill-thought-out "Hey, let's turn on the cameras and see what happens." No one who has her act together is probably going to want to do that, because not much quality is likely to come out of it.

- Always set a deadline for when their footage should be in your inbox. If two different people contact me about making a video with them, one who wants the footage by next Friday and the other who wants it "whenever I get a chance," it'll probably be a while before the second person gets anything from me.

- Sometimes it helps if your collab concept fits into a plot thread the other video maker has already commented on. When I asked Spricket24 and nalts to appear in my prank calling video, the two of them had a faux e-fight raging on YouTube, and I was able to work it into the script, making it more relevant and fun for both of them to participate.

- Never be demanding. You'll need to be proactive; people are generally busy and forget things. Sending occasional follow-up or reminder emails is okay. But, accept that it's also okay for whomever you're contacting to say "no" or to not reply at all. Not everyone wants to participate in the community or in collaboration videos. Always be polite when contacting people—don't burn bridges.

- It's bad form to say "Thanks in advance" in a business matter (and a collab video, even if no money is involved, is more or less business). Even if it's not, in your mind, you should conduct all interactions in a professional manner, even if you're light and breezy with the people while doing it. It will help build your reputation as someone who can be relied on—a "go-to" guy or gal—which is a great position to be in.

The reason you shouldn't thank people in advance is that it's assuming they'll say "yes," and that puts pressure on the person to do what you're asking. They might not have time or a desire to collab with you, and this makes it harder for them to politely say "no" if you thank them in advance. They may never want to work with you again if you put pressure on them. Sign your request email with "Thank you" or something like that, but don't sign it with anything like "I thank you in advance for doing this."

- Be willing to reciprocate. Don't get someone to collab on your video and then say you don't have time when they ask you later to collab on their video. Make time for them, if they've helped you.

Of course, you should keep in mind a few technical issues as well. Make sure you ask for file formats and compression codecs that your video-editing program can open. If you're on a Mac, you may not be able to use *.wmv* files, for instance. What about the aspect ratio? Videos look more cohesive if they remain in one aspect ratio—4:3 (full-screen) or 16:9 (wide-screen)—throughout. (See the "Aspect Ratios" sidebar for more about aspect ratios.) If these are details you're concerned with, make sure you mention them up front.

ASPECT RATIOS

The *aspect ratio* of your video refers to its width divided by its height. The most common aspect ratios are 4:3 (also known as *full-screen*) and 16:9 (also known as *wide-screen*). The majority of editing software can handle either format, and most can even mix the two formats within a single video, without you having to adjust anything. Each program handles output aspect ratios differently, but most of the time this is something you can set manually when you get ready to render. When working on a collaboration video, the project will feel more seamless if everyone involved is shooting at the same aspect ratio, so if this is important to you, mention it up front.

Finally, be kind in your editing. Some users will send more than one take so you can choose which best fits your scene. Others will pre-edit their material before sending it; you should respect those edits and not chop up the footage any further without asking them whether they're cool with it.

Large video files can be sent easily using host sites/programs such as YouSendIt, *www.yousendit.com* (URL 8.4) or Pando, *www.pando.com* (URL 8.5) if you don't have your own server (see the "Using YouSendIt" sidebar for more info).

USING YOUSENDIT

If you've never used or heard of YouSendIt, it is a website that lets you send any large file (up to 100 MB for free) to any email address. Most email addresses can't accept attachments that large, so, instead, YouSendIt saves the large file on its server and simply sends you and the recipient you designate a link to the file. It's very easy to use.

Simply enter the email address of the person you want the file to go to, a subject (title) for your email, and an optional message, and then select the file on your computer you want to send (see Figure 8-6). Once you have these basic pieces of information filled in, click Send It, and the file is uploaded from your computer to YouSendIt's server. YouSendIt will then dispatch an email automatically to both you and the recipient with a link to the file in the email message. It's really that easy to send collaboration clips to others.

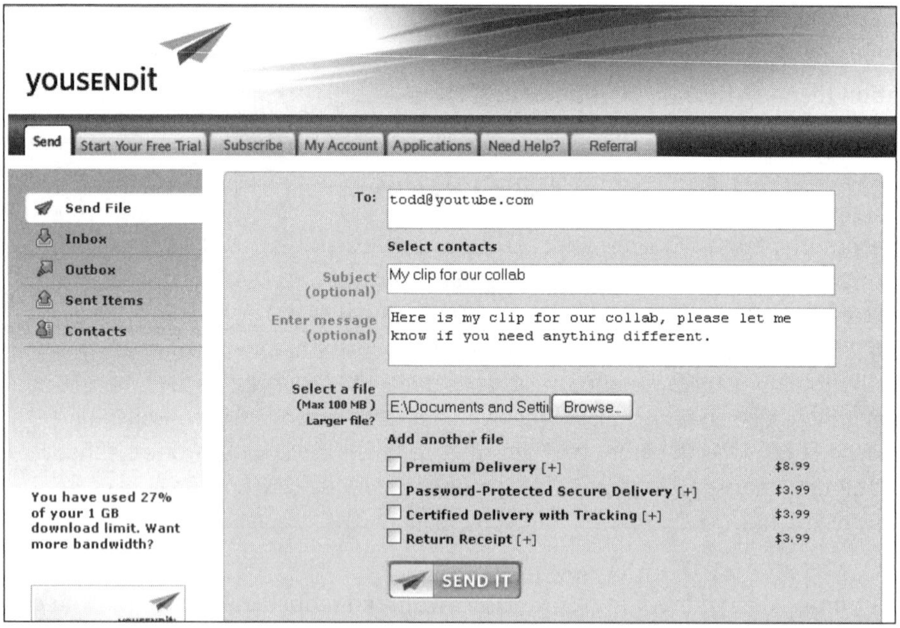

Figure 8-6. YouSendIt easy to use for sending large collaboration video clips to other YouTubers.

Remember to have fun with it. Collaborating is a great way to work with the viral video makers whose work you enjoy. Collaborations also expose your audience to new channels or personalities, and, in return, their subscribers will have the opportunity to see you.

Collaboration Channels

With collaboration videos proving to be fun and rewarding, YouTubers started to get ideas for collaboration *channels*. Collaboration channels are any YouTube channel shared by more than one user. Usually, these users make videos communicating to a particular audience of shared interests or simply to the other users on the channel. The vlogbrothers, John and Hank, held the earliest collaboration channel that I'm personally aware of, but the trend wasn't popularized until January 2008 when five female viewers of the vlogbrothers created their own home on YouTube to communicate to each other. Those viewers became known as the fiveawesomegirls (Figure 8-7). Their channel idea was unique; they would each take a weekday, vlog about their day, or answer questions the other girls had posed in hopes of making new friendships and strengthening old ones. (See the "Interview with fiveawesomegirls" sidebar.)

Figure 8-7. The five awesome girls of fiveawesomegirls

INTERVIEW WITH FIVEAWESOMEGIRLS

www.youtube.com/user/fiveawesomegirls (URL 8.6)

If you mixed *Brotherhood 2.0* with the *Sisterhood of the Traveling Pants*, and threw in a little *Harry Potter* and good genes, then you'd have the fiveawesomegirls.

Alan: Who are the fiveawesomegirls (5AG), and how did you meet?

Lauren: We're Kristina (italktosnakes, *www.youtube.com/italktosnakes*) (URL 8.7), Lauren (devilishlypure, *www.youtube.com/devilishlypure*) (URL 8.8), Kayley (owlssayhooot, *www.youtube.com/owlssayhooot*) (URL 8.9), Hayley (hayleyghoover, *www.youtube.com/hayleyghoover*) (URL 8.10), and Liane (lianeandthemusic, *www.youtube.com/lianeandthemusic*) (URL 8.11). We all met in one way or another through *Harry Potter*, YouTube, and each other.

Continued

Kristina: I'm kind of the hub of everyone...some of the girls knew each other pre-fiveawesomegirls, but I reached out and grabbed my very best fandom friends for the project and then crossed my fingers that everyone would get along.

Hayley: We're a collection of nerdy-yet-surprisingly-hot girls (readers, musicians, writers, actresses, and filmmakers) between the ages of 16 and 20. Each of us has a weekday on which we must post an approximately 4-minute video giving updates on our week, stating why the day is awesome, and sometimes completing themes, games, or truth-or-dare questions/challenges. The project started January 1, 2008, and will run for the entire year. We knew each others' names and faces from *Harry Potter* events and websites. Kristina was our mutual acquaintance, and she got us all together on Facebook with the idea.

Alan: What would someone tuning in for the first time expect to see over the course of a week?

Kayley: Generally, you'll just see five best friends talking to one another about what's going on in their lives. We all have similar, nerdy interests that we enjoy discussing as well. Some weeks there may be themes for our videos, which is always fun!

Hayley: Want a random sample? Well, Kristina once popped out of a cardboard box and pretended to be a hatching dragon baby; Lauren gave an inspiring speech about not being afraid to show the world the real you; Kayley pretended to sneeze every few seconds to go along with the "Seven Dwarfs" theme; I made a fast-motion montage of getting my hair highlighted; and Liane played a hilarious song about her crushes on Professor Lupin from *Harry Potter* and Jacob Black from *Twilight*.

Alan: Are there any rules?

Lauren: Each girl has a specific day of the week when she has to post a video to the channel before midnight (in her time zone). In every episode, we have to say why today is awesome. And occasionally we give out challenges or ask each other questions with the roll of a die. If one of us breaks a rule or deserves it, the other girls can pick a punishment for her to film.

Kristina: It feels like the rules are getting more and more relaxed. We used to be really serious about punishments and whatnot, but none of us has been that great about following through with the threat of punishments after about the six-month mark.

Alan: Do you feel that sharing the channel has helped you get to know each other better?

Lauren: Oh, definitely. I didn't know Hayley or Kayley at all when we started, and now I consider them some of my best friends. Kristina and I have been close ever since we met a few years back, and now we have even more in common. And Liane is a friend I've known for a little longer than we had the channel, and this has been a great way to get to know her. I live closest to her, so now we try to hang out as much as we can.

Kayley: Oh, definitely. I had only ever really talked to Kristina before this project. Now I have three other friends who make me smile just every single day. It's been fantastic these last six months getting close to all of them.

Kristina: 100 percent yes. If nothing else, even if no one had watched our channel but ourselves and oneawesomefanboy (hee hee, Jerry). I still think we completely found a magical connection with each other. It's actually impossible for us to put into words how much we all love each other, so don't bother asking. Just assume it's a whole lot.

Hayley: Without a doubt. Having watched their videos and followed their music, before the project started I thought all four of the other girls were way too cool to talk to me. Now I consider them some of my best friends. Kristina and Lauren came to my house while touring with their bands, and hugging them for the first time just felt so right. People don't understand how you can be this close to someone you've never touched or been in the same room with, but held at gunpoint I would swear that I know and love the fiveawesomegirls more than a lot of the people in my everyday life.

Alan: How do you feel about all of the spin-offs and imitators? Do any of you regularly watch any of the spin-offs?

Lauren: I think they're awesome, but I have to admit that the name copying is getting really annoying. We obviously didn't come up with the idea to share a channel, but I'm still kicking myself for being partially responsible for something like 758,947,586 [number]awesome[nouns] channels on YouTube. The spin-off I watch regularly is the wizrockateers. I knew most of them before they started the channel, and I admire their creative choice of a name.

Kayley: I am really happy that so many people seem to love the idea. It's quite an honor to just be a part of it all. I'll agree with Lauren, though, that it has gotten quite out of hand. I just try not to let the whole imitation deal get to me. I don't watch any of the spin-offs daily, no. It's very time-consuming! But I am subscribed to the guys' channel and watch it as much as I can, as well as vlogramen and agroupoffriends, which are mainly inspired by the vlogbrothers though did take a little inspiration, I think, from us.

Kristina: I watch the fiveawesomeguys pretty regularly, and every once in a while I'll tune into the 5awesomegays. Personally, I am just completely flattered that so many people liked our idea enough to want to do it too. The name copying doesn't even bother me; we had no idea it would take off like this, but I suppose it's our own fault for choosing a catchy name. When I see another collab channel that's clearly a spin-off of ours, it makes me kind of proud to know we started something that's affecting people.

Continued

Hayley: I'm going to be honest, here: I'm the one who had a problem with the spin-offs. I mean, at first it was flattering that we were inspiring other groups to be part of something this cool, but soon it became apparent that a lot of them were in it for popularity and it was sort of cheapening what our channel was all about. I didn't want to be grouped in with "all those fiveawesome channels" that just clogged up people's subscription boxes when what we had was a clever, original idea. Nowadays, however, it's become apparent which channels have something real going and which don't. The majority of the failed attempts fall into the latter category. I'll admit that it still infuriates me to be referred to as an imitation of the fiveawesomeguys, but I've come to accept that that isn't the guys' faults. They've respectfully provided links to our project and been very supportive in our times of trouble. I regularly watch the fiveawesomeguys and wizrockateers and will occasionally mosey over to the 5awesomegays and vlogramen.

Alan: Any downsides to sharing a single channel? Why do you feel so many of the spin-offs have failed?

Lauren: I don't know of a downside to sharing. I do have my own personal channel that I can post whatever I want on. The spin-off I watch has not failed, so I didn't even really know that so many of them were failing. My guess is that they started it for the novelty, because everyone else was doing it, not because they really wanted to maintain friendships with people who live far away. Projects like this require a lot of work, and they're difficult. If you don't have a strong connection with those beautiful four faces who are waiting to see a video about how your day has been, it can feel like a waste of time.

Kayley: No, there aren't any real downsides to it. I mean, I have a separate channel if I don't want to post something on 5AG, so there isn't much to worry about there. I actually wasn't aware many spin-offs had failed! That's too bad. Well, maybe it's because you have to be really dedicated to what you are doing when in a collab channel (or any time in life, really) and some people aren't willing to dedicate themselves to a channel. It is hard work, and sometimes it really isn't fun to have to record a video after a bad day—but we do, and we love it regardless.

Kristina: The only real problem with a shared channel is sometimes if there are decisions to make, there are so many of you that everyone just assumes someone else is going to make it. It's kind of like the bystander syndrome...everyone assumes someone else will think of a new idea for challenges or get a nice banner up even though you've been partnered for months and are still using one deemed "temporary." But these things are just details, and the positive aspects outweigh the minor negative by a long shot.

Hayley: My personal channel, hayleyghoover, is for the scripted comedy videos I make about once every two weeks, so it's really very different from 5AG, and I never feel like it's a burden to post only one vlog a week over there. One downside is that sometimes messages and comments go unread. Other than that, it's comforting to know that if you need your video description changed or something, you can just text one of the other girls and have it fixed. We've also helped rid each other of disgusting center screenshots and put ads on each others' videos from time to time. I think a lot of the spin-offs have failed because of lack of communication. The five of us do a lot of our bonding off-camera through Facebook and Skype and our phones. Hank and John of the vlogbrothers (our inspiration) grew up together, so it's not like they were trying to introduce themselves through video. I think if we were to just to be *us* without textual communication, we would be flat and eventually die out.

Alan: What's your favorite 5AG moment so far?

Lauren: I'm ridiculously fond of the time that Kristina and I were on tour and pretended that our car ran out of gas in front of Hayley's house, mysteriously. I tripped up her front steps, and the three of us fell into a pile of hysterical laughter.

Kayley: Oh, dear, how can I pick? One that doesn't involve me would be when Lauren and Kristina first met. That was a wonderful moment. But this last weekend, Kristina and I went to a YouTube gathering and got loads of great footage. I don't think we would have been there if it wasn't for 5AG, so that was really fun, and we met some amazing people. Oh! And surprising the other girls when we went to Forks was great, too. :)

Kristina: Meeting Hayley and knowing that I'd now met every single one of these wonderful girls (even though I know they all despise the fact that I'm the only one who has, ha ha). Also John Green judging the rap battle. He's our collective hero, so that was a great moment for all of us.

Liane: I think the raps have definitely been a highlight. It's nice to not take ourselves too seriously.

Hayley: Oh, gosh. Liane wrote a *gorgeous* song about one of our favorite books, John Green's *Looking for Alaska*, which she posted the day before we had our channel hacked by some jealous YouTube loser. When we finally got our channel back the next week, I loved that Liane's song was one of the most recent videos. It's just so... who we are.

fiveawesomeguys

Soon after the fiveawesomegirls created their channel, I was contacted by YouTube user charlieissocoollike. He proposed starting a spin-off called fiveawesomeguys. I loved the idea and quickly grew to love the fiveawesomegirls channel. After a few weeks of watching, I took Mondays; Charlie McDonnell (charlieissocoollike, *www.youtube.com/charlieissocoollike*) (URL 8.12) took Tuesdays; Alex Day (nerimon, *www.youtube.com/nerimon*) (URL 8.13) took Wednesdays; Todd Williams (Toddly00, *www.youtube.com/toddly00*) (URL 8.14) took Thursdays; and Johnny Durham (johnnydurham19, *www.youtube.com/johnnydurham19*) (URL 8.15) took Fridays.

My first video on the fiveawesomeguys channel—the first video on the channel—gave a shout-out to the fiveawesomegirls who had inspired us and explained what we hoped to accomplish with the channel. Seven months later (as of this writing), we're still vlogging to each other every weekday, only finding replacements if we happen to be seriously ill or without an Internet connection (both occurrences are, thankfully, few and far between). The fiveawesomeguys channel is the project I am most proud to be involved with on YouTube, and with an audience of almost 30,000 subscribers, we have a great time connecting with each other and having fun with our viewers.

After the fiveawesomeguys channel was created, numerous spin-offs, parodies, and response channels have been made following the fiveawesome____ formula that the girls created. So many channels were created, in fact, that the phenomenon was spoofed in a video by YouTube user JohnCocktoston called FiveAwesomeDipSh!ts.

Sharing a channel on YouTube certainly comes with an unavoidable amount of risk; the more people who have the password to a particular channel, the less secure that password is. A number of the collaboration channels on YouTube have been hacked via various password-phishing methods, so if you do choose to participate in or organize a collab channel, make sure everyone involved understands how sensitive your password is and understands how phishing happens so as to avoid it.

YouTube Groups

You can find YouTube groups (*http://youtube.com/groups_main*) (URL 8.16) on the Community tab that appears at the top of every YouTube page. Groups are free to join and even free to create. If you have a topic you'd like to discuss or share with numerous users, creating a YouTube group is a great way to gather everyone in one place. Groups allow numerous members to post videos and discussion threads (like a forum) in one place on YouTube (Figure 8-8).

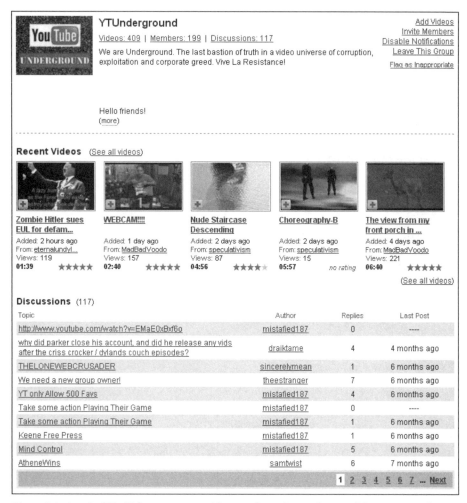

Figure 8-8. A typical YouTube group page with stats of members, videos, and discussions at the top and discussion threads at the bottom.

Groups exist for almost every topic. If you make videos about customizing cars, building instruments, or managing your finances, there's a group on YouTube for you. To join a group click the Join This Group link, at the top-right side of every group page.

After you're a member, add any uploaded video to the group, post new threads in the discussion section, or reply to questions and topics. To add a video to a group, click the Add Videos link, found in the upper-right corner of the group page (Figure 8-9). If you lose interest in the group or find a more relevant group, you can leave at any time by visiting the group page and clicking Leave This Group.

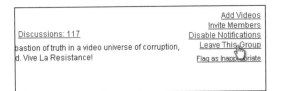

Figure 8-9. Most group interaction is done via the links in the upper-right corner of the group page.

Creating a New Group

If you can't find a specific group, create one. It's easy and free. After you've created your group, invite friends and other YouTubers with similar interests to add videos.

1. Go to your Account page (Figure 8-10) and Click Groups. This link is located on the lower-left corner of your Account page (Figure 8-11).

Figure 8-10. Go to your Account page.

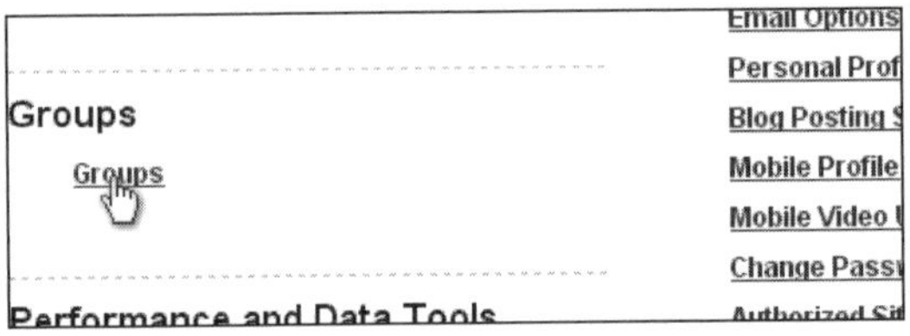

Figure 8-11. Click Groups.

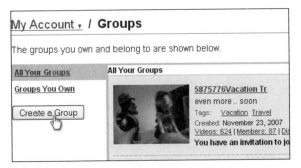

Figure 8-12. Click Create a Group.

2. Now you'll see a list of your groups. Click Create a Group. (Figure 8-12).

3. Enter your new group's information (Figure 8-13). What's your group called? What's the category? What keywords might someone search to find your group?

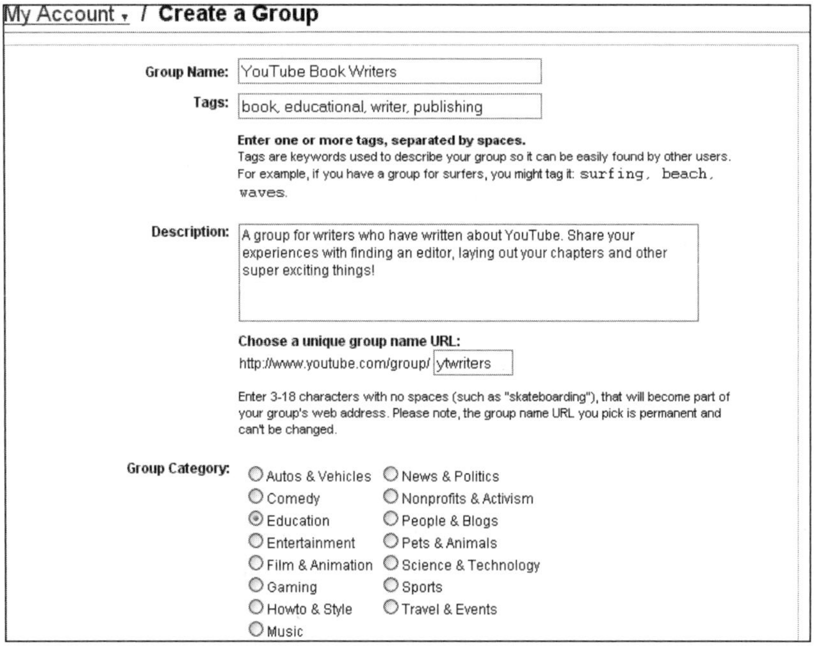

Figure 8-13. Add keywords and other details.

Figure 8-14. Click Create Group.

4. When you have finished filling out your group's particulars, click the Create Group button at the bottom (Figure 8-14).

5. You will then be taken to your newly created group page (Figure 8-15).

Group 'YouTube Book Writers' created.

Book Writers

Members: 1 | 0 Discussions

Edit Group De
Manage Vio
Add Vid

Figure 8-15. Your new group.

While you will find many groups created for specific hobbies or interests, a surprising number of groups are created to host contests. From time to time, YouTube, sometimes in conjunction with an advertiser, will run various video contests where YouTubers who create original videos have a chance to win a lot of money or other large prizes. One of the requirements of entering most contests is to join and post your video to the contest's official group.

Video Contests

Video contests (*http://youtube.com/contests_main*) (URL 8.17) offer YouTubers a chance to win prizes (cash, trips, or other expensive items) with their creative videos. You can find the contests on the Community tab that appears at the top of every YouTube page (Figure 8-16). Each contest has its own set or rules and requirements

Figure 8-16. YouTube's contests home page.

for videos. Some are humorous 30-second commercial contests; others are vlogging contests about a particular subject or news event.

Most contests are sponsored by large corporations such as automobile or soft drink companies; however, occasionally YouTube will sponsor its own event. One of the most popular contests is YouTube's annual Sketchies contest. The Sketchies contest calls on all the YouTube comedy channels to create original, humorous 3-minute shorts that incorporate various items or phrases assigned by YouTube. The first Sketchies contest was introduced by the Lonely Island group (featuring *Saturday Night Live*'s Andy Samberg), while last year's contest was introduced by past year's winners Awkward-Pictures and LisaNova.

If you're short on video ideas, the imposed rules and guidelines of the various contests might be a good way to spark your creativity and perhaps result in a bit of reward.

YourTubeAdvocate

The YourTubeAdvocate channel is unique. The channel changes hands every month, from one YouTuber to another. The YourTubeAdvocate for any given month attempts to report any site updates to viewers, while also taking users' issues to YouTube. While not an official project, YouTube has, in the past, responded to numerous advocates' emails.

Picture the YourTubeAdvocate as somewhere between ambassador, senator, public relations person, and neighborhood watch.

It is sometimes difficult to contact YouTube staff, and most emails are responded to with automatically generated messages, which often do not really address the concern or issue you wrote in with. Other email responses point you to the Help Center (*http://help.youtube.com/support/youtube/bin/static.py?page=start.cs&hl=en_US*) (URL 8.18). Therefore, having the YourTubeAdvocate channel as one additional avenue to potentially reach YouTube is really helpful.

The channel was created by YouTuber JustAllen (*www.youtube.com/JustAllen*) (URL 8.19) and has changed hands more than a dozen times as of this writing (Figure 8-17). I held the position for the month of November 2007. During my one-month term, I made eight advocate videos, including one, which YouTube later featured, about celebrities like Oprah invading what we felt was our YouTube.

Figure 8-17. achampag as the YourTubeAdvocate for May 2008.

If you're looking to be more involved in the community, feel you could contribute educational or interesting videos for a solid month, and have invested a bit of time in your channel already, emailing the current YourTubeAdvocate and letting him or her know you're interested is a great first step. Even though the advocate position doesn't pay anything and you'll deal with more emails in one month than you have in the past year, it is a completely rewarding experience.

Trolls

Not every comment or email you receive will be positive or even constructive (on YouTube, as well as everywhere else in life). You will receive random hateful comments from time to time. These seemingly random comments will have no basis or substance and will usually be very impersonal. Hate comments typically take the form of "You're fat," "Wow, you're boring," or "You're gay," and the hater posting the comment will usually spell *your/you're* incorrectly.

However, a *troll* is a hater who singles you out, or targets you personally, repeatedly. Trolls have objectives. A troll may want to irritate and upset you so much as to run you off YouTube. They build their low self-esteem by doing this. Resist the urge to engage or "feed" any potential trolls. Simply ignore them.

Trolls will get bored when they see that their tactics don't provoke any response from you at all, and they will move on to someone they can upset. I've personally seen a number of my friends lose accounts or become extremely upset over trolls, but learning to deal with hate and detractors is a skill that will serve you across all the Internet, not just on YouTube.

And don't *be* a troll, even a little baby troll, which is possible even if you don't think you're doing it. Avoid ad hominem replies and comments. (*ad hominem* is Latin for "attacking the man" and means going after the person, rather than just making a comment about something you don't like in their art.) I generally avoid making comments that are negative anyway, unless it's constructive, like "I like your video, but I had trouble hearing what you were saying. Here's how to get better audio, if you're interested...."

As the late, great Liza Matlack (*www.youtube.com/watch?v=dRf6DZYY6dg*) (URL 8.20) said in Michael's film *D.I.Y or DIE: How to Survive as an Independent Artist*, "Mean-spirited criticism doesn't ever come from someone who's doing valuable art themselves. Almost everyone who does valuable art, never themselves, *never*, delivers mean-spirited criticism because they know how to get from point A to point Z...and they know you need to go through B, C, D, the whole shebang, to get to where they are, and they understand that, and there's no need to criticize."

This is important: When people post mean stuff on your videos, they're jealous. They're usually very young teens with no social skills and unfulfilling lives posting from the safety of their parent's paneled-basement rec rooms, hiding behind the anonymity of the Internet. They feel safe to say stuff online that would get them punched in the face in person. That's why they say so much of it online. They're cowards, and they're jealous because they're too stupid and untalented to do something cool. For people like that, it's easier to destroy than to create.

Don't tell them any of this; it's a waste of time, and "winning" an argument with them is a very empty victory and will waste time you could be using to create more cool stuff. Just delete the comment, block them, and move on. They'll go find someone else to bug. They need to feed off of someone who will argue back, because they don't have a life of their own. Trolls are serenity vampires.

All of this "Don't feed the trolls" advice is good to heed on the 'Tube, as well as on any other web community but also, and especially, in real life.

Your Niche

You have a lot to gain from being an active member of the YouTube community. You can make some great friends, take part in some really creative and organic projects, or find a dedicated audience for your work. But being active in the community is not for everyone. It can be time-consuming. It can even be frustrating when friends of yours get into arguments or decide to move on to a different website. Keep things balanced; don't let YouTube become your life. Find a circle of friends with which you feel comfortable, and support others who make videos you enjoy. They will be the first to support your projects.

And remember that it will take time. Just like your first day of high school, you're not likely to know anyone yet, and it will take a few days or weeks before you find people you like spending time with. Don't try to fake it. Don't try to inflate your views or subscribers by cheating. Let these things build naturally.

In the next chapter, we will describe some of the tactics people have used in the past to cheat so you can identify them in others, as well as avoid using these tactics yourself.

9

Hacking the System: How to Cheat (and Why You Shouldn't)

Alan Lastufka

A Word of Warning

Every tip, trick, strategy, and suggestion offered elsewhere in the book is 100 percent safe and within the terms of use and community guidelines published by YouTube. This chapter is the only exception to that statement. While the other information in the book may be considered "selling out"—or even whoring yourself—it is all legitimate promotion of your channel. Do not—I repeat, *do not*—attempt the following gimmicks to build up your channel. This information is included for educational purposes only, because, as with any system, YouTube can be cheated and gamed.

Each of these cheats is a surefire way to lose your channel. You may see others using them, you may even see others get away with them, but please resist the urge. Should you attempt any of the following and your channel be suspended as a result, there will be little you can do to get it back; you will have knowingly violated YouTube's user policies.

I'm just telling you what some people do so you know a little more about what's "under the hood" of the car that is YouTube and how some people "hot rod" it against the rules (and often get burned in the process) so you'll understand why some people may seem for a minute to be more popular than they probably should be. Also, a few of these "tricks" might be something you'd come up with on your own and not realize that it's something you're not supposed to do and could lose your account over. So, I'll just give you the heads up on all the "hacks."

(We've decided *not* to give the URLs of the cheating software. If you want to find them on your own, that's at your own risk.)

Spam Bots

Spam bots are software programs that will repeat a certain action, such as leaving comments or sending friend requests, over and over...and over and over. TellyAdder is probably the best-known spam bot directed at YouTube spamming. In fact, its own website touts it as "YouTube Viral Marketing Software." TellyAdder (and the majority of other spam bot software) works by building a database of usernames and video links; you can even build multiple databases for different actions (Figure 9-1).

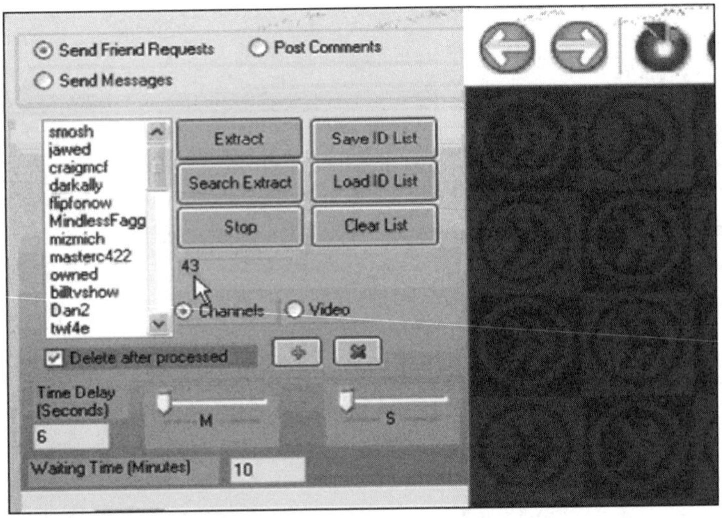

Figure 9-1. TellyAdder's default screen when building a database.

Let's say you want to leave a profile comment on all your subscribers' channels. This would take some time to do manually; even copying and pasting the same message over and over would take hours if you have more than 100 subscribers. However, using TellyAdder, you could build a database of your subscribers' channel links, which TellyAdder can do relatively quickly by scanning YouTube pages and grabbing all channel links from the pages. Then you could simply reload that database every time you wanted to leave comments for all your subscribers.

Using this process to promote outside websites used to be abused a lot by numerous YouTubers. Most of these users were promoting adult websites or other commercial services; they weren't actual Tubers who uploaded videos. The spamming became such an annoyance that YouTube users LisaNova and TheDiamondFactory made a 10-minute video spotlighting the problem, which went on to be featured on the front page of YouTube and to earn almost 2.5 million views (read more about their project in Chapter 7).

YouTube received so many complaints it eventually initiated a CAPTCHA code system (Figure 9-2). After leaving four comments within a certain amount of time (most likely less than 15 minutes), you are asked to enter a string of characters to prove you're a human leaving comments, not a bot. TellyAdder, and other bot programs, can't read the CAPTCHA codes from the screen and therefore are no longer effective because they cannot leave more than four comments in any 15-minute period.

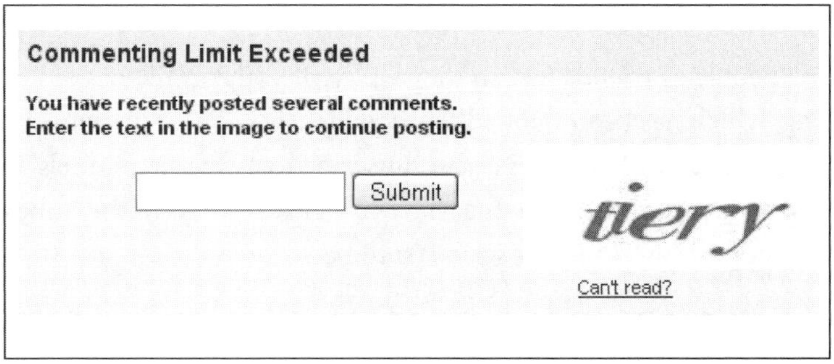

Commenting Limit Exceeded

You have recently posted several comments.
Enter the text in the image to continue posting.

Submit

tiery

Can't read?

Figure 9-2. YouTube requires you to identify a CAPTCHA code before leaving more than four comments in a row.

Ever since the CAPTCHA program began on YouTube, we've all noticed an impressive drop in comment and channel spam messages, although you still may receive some. Most spammers now seem to focus on the Most Viewed videos of the day and leave smaller, less visited channels alone.

Related to spam bots are *friend bots*. Friend bots accomplish the same goal as spam bots, in that they automate a repetitive task, in this instance, adding friends on YouTube. This practice doesn't usually net as big a result as comment spamming, but it is still useful to those who like to send out lots of bulletins. Also, adding numerous people as friends will typically trick those people into visiting your profile under the assumption that you added them because you liked their videos. They may even subscribe to you under this assumption.

Spam bots have one semilegitimate use, and that is automating posting "thank you" messages to many users who've commented on one of your videos. You'll have to type in the CAPTCHA every four comments, but it can still save you time. It's kind of cheesy to do this, and it isn't going to win you many real friends, but some people do it. I can't find anything in the terms of use that directly prohibit it, but read the YouTube terms of use

http://youtube.com/t/terms (URL 9.1)

before doing this (or anything that might seem like a "hack"), because the terms of use are always subject to change. (They're called *terms of use* on YouTube but *terms of service* on most other sites.) Terms of service are always a work in progress on any site, because new issues arise that have to be dealt with by the site administrators.

iSub Network and Sub4Sub

The practice of adding people as friends just so they will visit your profile has led to a new form of cheating, which has been dubbed the *iSub network*. The iSub network refers to the unorganized group of YouTubers who subscribe to as many channels as they possibly can (usually favoring channels that have few subscribers) in the hopes that those channels will appreciate it and subscribe (*sub*) back.

When a user with very few subscribers receives a new one, they typically visit that new subscriber's channel and watch their videos. Most of the time, this person will subscribe back to their new subscriber. The original user never intended on watching or interacting with this other person, but just got a subscriber bump out of the exchange. Take that instance and multiply it by thousands of subscriptions, and being an iSubber can help rocket your channel up the Most Subscribed list.

It used to be easy to tell who was running this scheme because YouTube had no real way of hiding how many people you were subscribed to. It can be argued that subscribing to more than about 500 channels means you will never have the time to watch all the videos posted by those channels. Some argue YouTube should cap your maximum subscriptions at 1,000. However, in the spring of 2008, YouTube did the opposite and actually created a way for users to hide the number of subscriptions and friends they had, effectively allowing the iSub network free rein to continue their practices without anyone being able to call them on it.

Similar to those who subscribe to everyone, secretly hoping those users will subscribe back, we have the *Sub4Sub* users. Sub4Sub users blatantly tell you they will subscribe to you if you subscribe to them. They leave comments encouraging everyone to subscribe and add them as friends, promising to do the same in return. And many of these users gain hundreds of subscribers with this method.

Despite its perceived success, both of these gimmicks are good only for short-term number boosting and will not build you a solid audience. Most Sub4Subbers don't watch videos, or they use a second account to view videos where they subscribe only to the users they actually want to watch. These gimmicks are designed to boost their numbers and won't benefit your channel. In fact, participating in these deceptive practices and begging for subscribers will only hurt your channel should you be exposed.

E-begging

Technically, *e-begging* isn't really a form of cheating, but it seems just as dirty to most users as the other cheating tactics discussed here, so I believe it fits. E-begging (a term claimed to be coined by utubedrama.com's Trevor Rieger, who you'll learn more about later in this chapter) is the act of asking for money from your online viewers. The majority of Web 2.0 sites do not pay their content creators (YouTube is one of the few exceptions; you'll learn more about YouTube's Partnership Program in Chapter 11). Therefore, in my opinion, as some users become more popular, they tend to treat their accounts more as a job than a hobby. This leads to some feeling they're "owed" for all the entertainment they've provided you with, so they ask for a donation.

A typical e-begging video is serious in tone, with an artificially humble YouTuber explaining some recent crisis they've had to endure (for example, a family member becoming ill and being out of work or their computer dying on them so they can't make regular videos). Once the crisis story is over, they will typically mention a way for you, the viewer, to send them money to help them through this tough time. While a small percentage of these videos may be honest and come from a truly desperate place, the majority has been proven to be fake or exaggerated.

One YouTuber went so far as to claim his younger brother was dying. This YouTuber claimed he needed your money to help his little brother. However, not only was his little brother not ill, his little brother didn't even exist. The YouTuber had completely made up a sick, dying sibling to ask for donations. This YouTuber has since removed his e-begging video from YouTube, so unfortunately, I can't link you to it.

Autorefreshing

Autorefreshing is the oldest trick in the book. I touched on it a bit in Chapter 7, but it's important enough to be mentioned again in this chapter. Autorefreshing involves using an Internet browser plug-in to refresh any page (like a YouTube video page) automatically at a preset time interval (Figure 9-3). For example, if I set an autorefresher to reload my video page every 5 seconds, that would produce 12 views per minute, or 720 views per hour. YouTube has claimed to take steps to correct this; it tracks which IP addresses are viewing which pages, but the cheat still works, and it will still rocket your video to the Most Viewed lists if used often.

Figure 9-3. Autorefreshers are easy to install and use, but they won't help you build an audience.

Users who autorefresh are easy to spot; typically, their video will have an out-of-control views-to-comments ratio. If a video has 100,000 views but only 12 comments, you know something is wrong. As with most of the cheating methods discussed here, autorefreshing is a short-term fix. It will inflate your view count, but it won't help you build a solid audience.

Thumbnail Cheating

A *thumbnail*, in this sense, is the tiny image that shows up on any YouTube page to the left of the title of any video. Thumbnail cheating is the act of using a misleading or blatantly sexual image as your thumbnail frame. Each video has a thumbnail image associated with it—one that is taken automatically from a single frame in the video. YouTube allows you to choose between three points in your video to act as the video's thumbnail. The first choice is the 25 percent point, the default is the 50 percent point, and the third is the 75 percent point. Knowing these percentages, video creators, while editing their videos, are able to insert images that they want to appear as the thumbnail. Some users will insert appropriate thumbnails that do relate to their video's content, but other users simply choose to insert an image of a scantily clad young woman or a close-up of her cleavage.

Along with the video's title, the thumbnail is usually the biggest factor in determining whether viewers will watch your video. It has been proven many times that using a sexually suggestive image as your thumbnail will result in thousands or tens of thousands more views on a single video than a natural thumbnail from the video would produce.

YouTube is clear that it does not condone this practice. The community guidelines on YouTube, *www.youtube.com/t/community_guidelines* (URL 9.2), state the following:

"Do not create misleading descriptions, tags, titles or thumbnails in order to increase views."

Even with this guideline in place, many YouTubers still "game the system" by using misleading thumbnails. When flagged, YouTube has removed or unmonetized videos using such techniques. But this guideline is largely ignored. The YouTube administrators have limited time to police a lot of YouTube real estate, and there are much more serious violations to worry about. Using a misleading thumbnail on occasion, especially for comedic purposes, probably won't get you too much flack, but repeated use or abuse will most likely upset the majority of your subscribers.

Tag Loading

Loading your tags (tags are also referred to as *keywords*) with unrelated words works much the same way as using a misleading thumbnail. You assign text tags to your video when you upload it, and you may edit the tags at any time. Because the search function can't determine what your video is about without additional data, these tags help place videos within search results. However, loading your tags with unrelated but popular search terms, such as *nude, sex, xbox 360,* and so on, can place your video in search results where it doesn't belong. It is misleading, and while it may bring in a few extra views, it will frustrate users searching for videos about a particular subject and finding your unrelated video instead.

Tag loading undermines the value of any site. It makes one of the coolest things about the Internet and computers (the ability to search things) not work as well. Doing this is selfish and really antithetical to the idea of being a member of the community of YouTube. It will make you look bad.

Tags should include words relevant to your video and to your channel name. Good tags are general search terms that are relevant and help people find your video.

Sock Puppets

A *sock puppet* is a user's secondary channel, usually used to mask the identity of the user. Sock puppets are usually good for only two situations:

- **Leaving anonymous comments**: If you feel strongly about a personal opinion, one you may not want to share with your general viewers, leaving a comment about that opinion under your own username would be risky. However, if you were to register a second account, a sock puppet, under a name no one could associate with you, you'd be able to speak freely. You may also want to leave anonymous comments if those comments would reflect negatively on you, such as when you're vlog warring or flaming another user. (This type of behavior is not encouraged, and I consider it a gross waste of time. Many people feel this way.) Using a sock puppet to make fun of someone, or to "drop their docs," will keep your regular username out of the drama.

- **Artificially inflating your comment, view, or subscriber count**: The comments you leave on your own video do not count toward the comment total. Therefore, some users use secondary channels, or sock puppets, to reply to comments or leave additional comments on their videos. Sometimes people create these sock accounts for the sole purpose of subscribing to their main channel, thereby bumping their subscriber count up with ghost accounts. Occasionally, these sock puppets are known to be used by the user; maybe it's even a secondary channel that they use for posting video responses or uploading test clips, but most of the time they are anonymous.

Utubedrama.com and Trevor Rieger

Celebrities have the supermarket tabloids running exaggerated or dramatized tales of their every move, misquoting their every step. YouTubers have Trevor Rieger and his "news" site, *www.utubedrama.com* (URL 9.4), as shown in Figure 9-4.

Figure 9-4. Utubedrama.com's Trevor Rieger gives well-known YouTubers their own page on his site; here's mine.

Trevor began on YouTube with the username utubedrama. He made videos in which he exposed high- and low-profile YouTubers for cheating, "gaming the system," or participating in other unfair practices. He found users who autorefreshed, and he grabbed screenshots of their behavior over time to present "evidence" of their actions. He looked through subscriber lists and shined his spotlight on those creating ghost accounts and sock puppets just to subscribe to themselves. He entertained a few people, got a few people removed from the site for breaking the terms of use, and made more than one person cry and threaten to leave YouTube after they were exposed.

DROPPING DOCS

Dropping docs is the act of posting someone's personal information on the Internet. It can be their legal name (if they use a screen name), address, telephone number, or any other piece of personal information. This is usually done by **haters**, users who take pride in and have fun disrupting the lives of others. Sometimes the numbers of the user's employer, friends, and family members are posted instead, and other haters are encouraged to call and harass these people. Most consider dropping docs a serious violation of trust and netiquette. It clearly violates YouTube's community guidelines:

"There is zero tolerance for predatory behavior, stalking, threats, harassment, invading privacy, or the revealing of other members' personal information. Anyone caught doing these things may be permanently banned from YouTube."

http://youtube.com/t/community_guidelines (URL 9.3)

Dropping docs is a violation of the terms of service of almost every social networking site in the world. In some cases (resulting in harassment or physical injury of the people being **outed**), it results in criminal prosecution. You could go to jail. Don't do it.

Eventually, Trevor upset enough of the right people, and his account was suspended. Trevor now continues exposing cheaters and YouTube celebrities on his own server. Utubedrama.com collects each day's activities from cheaters, or users gaming the system, typically by using one or more of the gimmicks listed in this chapter. The site also embeds a few key videos from the last few days, usually of e-beggars or people claiming they're leaving the site because of this or that.

The site is entertaining, even if most of what it "reports" seems to be taken out of context or tends to perhaps be misquoted. Trevor claims most of his information comes from other YouTube users ratting on each other and also boasts that he will keep all emails and messages in the strictest of confidence. When you're first starting out, you probably don't want to end up on one of Trevor's drama reports—that usually means you're doing something you shouldn't be. However, as your channel grows and you gain a few thousand subscribers, you'll become a target for people like Trevor, and at that point, you'll be able to laugh off most of what he reports. I certainly did.

This Account Is Suspended

YouTube can suspend your channel at any time for any reason (Figure 9-5). In fact, the site doesn't even need a reason to suspend you. However, if you keep your account in good standing, by avoiding the gimmicks in this chapter while also uploading only original videos, you should have little to worry about.

> This account is suspended.

Figure 9-5. The only thing left of an account after it has been suspended.

If you are suspended, YouTube will send you an email explaining why. Nine times out of ten when my friends have been suspended, it has been because of copyright claims. Don't use music that doesn't belong to you, and don't upload videos you didn't make. That sounds easy enough to me. Should you be suspended, however, here are a few tips that might help you get your channel back:

- Do not create a new channel and upload a video cursing out YouTube staff members for suspending you. I know it's tempting. I know you're upset and angry and feel wronged. But this will only hurt your chances of getting your channel back.

- Do not ask all your friends to spam or email bomb any of the YouTube editors with messages demanding you be reinstated. Again, this will only hurt your chances of coming back.

- Do calmly and politely email YouTube and, if you haven't been informed already, ask why you were suspended. If you disagree with the reason, calmly respond with your side of the story. If you were at fault, your best option is to sincerely apologize and throw yourself at the mercy of the YouTube editors. If you haven't made many negative waves on the site in the past and you're handling your suspension with a certain maturity, you might get your account reinstated. I have seen numerous accounts reinstated when people follow these simple guidelines.

All that said, YouTube is very clear and firm in its stance that once you are suspended, you are suspended for life. You are not allowed to create a new account and return.

If you follow the rules most of the time and you're respectful to the staff, again, you should have nothing to worry about.

I hope you use the information in this chapter to recognize cheaters and learn from all of their mistakes. It has been said that a wise person learns from the mistakes of others rather than having to make all the mistakes himself. I encourage you to refer to this chapter first before taking any action if you are suspended. But mostly, I hope you never have to refer back to this chapter, because you've quickly internalized the simple steps to harvest the joys of being a contributing part of the YouTube community, rather than taking your chances as a cheater only looking out for yourself. Happy honest 'Tubing!

Ethical Hacking

Michael W. Dean

The Internet was invented during the late 60s and early 70s by the U.S. Army as a way to network computers.

The World Wide Web as we know it, the graphical part of the Internet, started in 1992 as a way for physicists at a lab in Switzerland to get their computers, which all had different operating systems, to talk to each other. It was invented by one person, Tim Berners-Lee; see *http://en.wikipedia.org/wiki/Tim_Berners_Lee* (URL 9.5).

In 1997, I wrote Tim an email thanking him for inventing the Web. It's the only fan letter I've ever written. He wrote back. Cool guy.

The Web quickly spread to other research facilities and universities around the globe, began trickling down to nonacademics in about 1995, and became the media darling—the "next big thing"—in about 1997. I got online in 1996. I had a computer in 1991 but used it only for word processing. I watched the Web grow from something used largely by academics to something that bankers who'd never touched a computer were investing millions of dollars into.

Lots of money was quickly poured into crazy Internet business plans. The joke, and it wasn't far from the truth, was that companies with business plans written on napkins were getting millions of dollars in funding. This period saw the rise of web services that did things like delivering gourmet dog food from orders over the Internet, and startups that never shipped a unit bought commercial time during the Super Bowl. Many services promised to "expedite interactive partnerships" and "streamline real-time web readiness." No one knew what this meant, including the bankers who were writing the checks. There were companies that burned money on their launch party, like Pixelon, who claimed to have a video-sharing service sort of like what YouTube eventually became. Pixelon raised millions and hired The Who to play their private parties, but it seems they had no real technology to sell. Investors were horrified to find out later that their CEO had been previously arrested on a stock-fraud conviction. See *www.wired.com/techbiz/media/news/2000/05/36243* (URL 9.6) for more.

It was a new gold rush, and the epicenter was San Francisco, the same place where the first gold rush had been in 1849. I was making $30 an hour in 1998, working as a writer for several companies that went under before they ever went public.

Lots of these business ideas were silly, and lots of people lost a lot of money. In about 1999, the whole dot-com bubble burst. People and companies licked their wounds for a year or so and then went back to the drawing board, retooled the ideas, and figured out what worked and what didn't; in about 2000–2001 Web 2.0 was born. Web 2.0 wasn't about businesses selling directly to people; it was about people communicating their thoughts with other people. YouTube is Web 2.0. So is blogging, Flickr, MySpace, LiveJournal, Facebook, and many other newish companies that are thriving at the moment.

The Hackers

Meanwhile, as all this commercial futzing was going on, and even before it, there was a mostly noncommercial spirit of invention driving it all. That was the hackers. Until about 1996, when the press got it mixed up, the term *hacker* didn't mean a bad guy. *Hacking* didn't mean destroying computer systems or stealing data, such as credit card numbers, like it does now. The term for that was *cracking* (as in *safe cracking*, the term for breaking into a bank's physical safe), and the people who did it were *crackers*. Back in the day, and even now, there are ethical hackers (now often called *white-hat* hackers, as opposed to the bad guys, who are *black-hat* hackers) who drove the computer world and often were the people inventing the technology used by everyone else.

 Ethical hacking, while "ethical" in my mind, should never be done if it breaks any law or violates any terms of service of any site or system.

One story that has been pretty much verified as true by people who were there (and covered in the excellent movie on the history of the home computer business *Pirates of Silicon Valley*)

www.imdb.com/title/tt0168122/ (URL 9.7)

is that the two guys who started Apple Computer (the first company that made computers for the average person, rather than just for businesses), Steve Jobs and Steve Wozniak, were inspired to make computers by first making *blue boxes*, little electronic devices that allowed you to make free long-distance phone calls. They were inspired to do this by their friend John Draper (aka Cap'n Crunch)

http://en.wikipedia.org/wiki/John_Draper (URL 9.8)

who discovered that at the time (the early 70s) you could get free long-distance phone calls by whistling into a phone with a plastic whistle that came free as a prize in Cap'n Crunch cereal boxes. John Draper later invented the first-ever commercially available word-processing program, EasyWriter.

These guys weren't so interested in cheating the phone company. Their motivation wasn't crime; it was simply seeing whether it could be done and figuring out ways to do it. The home computer business was started and driven by people who definitely had a hacker ethic about them. This hacker curiosity led them to perfect technologies that transformed the world, often working from their bedrooms or garages. They were not all businessmen (although Steve Jobs certainly was, albeit a businessman who liked to go to work in bare feet while running a billion-dollar company). Steve Wozniak said on his blog,

www.woz.com/letters/pirates/33.html (URL 9.9).

"Good luck with your own dream. My dream was actually just to have a computer some day. If I'd imagined that it meant starting a company to sell them, I probably would have avoided the whole thing."

The ethical hacker idea exists today, driving some of the most stunning innovation in computing and on the Web. Ethical hackers do not just use the Web; they often build the Web.

Ethical hackers are the people who just want to "look around," who like to take things apart and put them back together, and who maybe make them better than they were in the process. It's a type of mind-set that can get you kicked off a site like YouTube. (With good reason; there are too many people on YouTube for the admins to figure out who's a white hat, who's a black hat, and who's just trying to selfishly artificially inflate their views, so they pretty much ban any hacking-type behavior.) But if you have a lot of curiosity about things and how they work, are good at figuring them out and even improving them, and want to devote that curiosity to good, never to evil, it can lead to bigger and better things.

INTERVIEW WITH JOSEPH MATHENY

In that spirit, here's a short interview with my very talented and very employed friend Joseph Matheny, a social media technology consultant, who is an ethical hacker.

Joseph Matheny, www.linkedin.com/in/jmatheny (URL 9.10).

Michael: Can you give a brief bio of yourself, including types of jobs you've had?

Joseph: I've held positions at Macromedia, Netscape, Adobe, Spruce DVD, Intronet-works, and MediaTrust. I have served as a webmaster, IT manager, programmer, product manager, and chief technology officer (CTO). I was on the Haiku branch of the DVD standards committee and was one of the originators of the art form known as *alternate reality gaming* (ARG). I also wrote one of the most widely republished white papers on DVD technology, "Why DVD? A Meat and Potatoes Guide for the Uninitiated."

[http://search.barnesandnoble.com/Why-Dvd/Joseph-Matheny/e/9780967489001 (URL 9.11)]

Michael: Tell me the story of how you helped turn the word *Photoshop* into a verb.

Joseph: This must have been around 1996, early in the Internet boom years. Adobe had been a print technology company and was getting into the Internet but had some catching up to do. Back then, people in the Internet Products Division referred to me as the God of the Web, because I had been working online since the 80s with BBS and Usenet. I had become the guy to go to with questions regarding Internet technology and culture. Someone sent a lawyer down from Legal one day to ask me some questions about the "Juarez" problem.

Continued

I replied, "Juarez? What's going on in Mexico? I thought Hong Kong was the problem; that's where most of the pirated discs are coming from."

The lawyer paused, and then said, "No, Juarez. W-A-R-E-Z."

"Oh, Warez!" I said. "You mean pirated software?"

"Yes," he replied, "so it's pronounced 'Where-ez'?"

So, we talked for a while, and he told me his concern that sales of applications like Photoshop were being affected by the Warez scene, and I explained that it wasn't really a problem, not the way I saw it. "In fact," I told him, "This is actually a boon. Look at it this way. The kids who are trading ZIP files of Photoshop distributions would have never bought it in the first place. In fact, I can tell you from firsthand experience that people have drives full of these things; they collect them, trade them like pogs, but rarely install them. If they do install them, it's to crank out some KPT-enhanced buttons for their Geocities website, and then it lies fallow."

[KPT = Kai's Power Tools, a set of plug-ins for Photoshop that easily allowed you to make really cool-looking buttons for websites or cool graphics. Check out their site: *http://tinyurl.com/66ylkd* (URL 9.12).]

I then explained how I thought that this helped with the word-of-mouth marketing of the product. Back then, Photoshop was a verb among graphic designers, but the term had not quite crossed over into the mainstream. I spent a lot of time convincing Legal that they should ignore "Warez kids" and focus on commercial piracy outfits like those in Hong Kong at the time and that the benefit is that all these Internet denizens would be chattering about "Photoshopping" things and help build the mind share in the noosphere. They agreed (at the time) to focus on the bigger fish. It was about a year or so later that I first heard the term ***Photoshopped*** used in a prime-time TV show. Later, Adobe got very aggressive about piracy, but that was yet to come.

[Noosphere = the collective consciousness. To learn about "Noosphere," go to: *http://en.wikipedia.org/wiki/Noosphere* (URL 9.13).]

Michael: Tell me your thoughts on "ethical hacking." Should people be able to "poke under the hood" of a site or service in a way that might be a violation of terms of service, as long as nothing is disturbed and no one is harmed? What are the advantages of doing so?

Joseph: I think that poking under the hood is A-OK as long as no data is stolen or harmed. The ethical thing to do is to leave the admin a calling card—a little text note telling them what you did and how you achieved the compromise—and maybe give them some help if you know how to secure things in the future. All you're doing is helping to secure the system for users and hopefully help the admin avoid a malicious attack. I strongly dislike defacers.

Michael: How have you "lived outside society's expectations" while charting your own course and making your mark?

Joseph: Ah, I don't want to think about that too much. I think everyone should follow their bliss and not be too self-conscious, lest they lose contact with the muse of nonlinear dynamics. Look forward. Only look back long enough to gather lessons and materials necessary for your next iteration of being.

Michael: What do you think of social networking in general and YouTube specifically?

Joseph: Like any new application of technology, we've had to live through the hype. What is social networking? We do it every day in meatspace. We've been doing it online from day one. BBS, Usenet, forums—they are all social networks. The only thing we're seeing now is some new applications of existing technologies for social purposes. These have both a light side and a dark side. Stalking is a problem, as is fraud like phishing. However, I've noticed that applications like Facebook (used in moderation) and Twitter help me stay current with people that I might otherwise lose touch with. In one way, I guess using software to stay in touch with people is a symptom of our busy lifestyles.

I like the sheer diversity that sites like YouTube have enabled in the video space. I'm looking forward to the next generation of semantic applications applied to the Web. I expect sites like YouTube to be used as the content bases for the first experiments in truly semantic video applications.

[*Meatspace* = the real world; people-to-people, face-to-face communication, as opposed to interactions over the Internet. See *http://en.wikipedia.org/wiki/Meatspace* (URL 9.14).]

[*Semantic applications* = programs that embed metadata, or "data about data." *See http://en.wikipedia.org/wiki/Metadata* (URL 9.15).]

Summary

So, there you have it, gentle reader. That was a crash course in what you can do, why you shouldn't, and how natural curiosity, combined with a shining internal moral compass, can help you get rich and change the world. Maybe. Or at least do cool stuff.

With that under your belt, you can now continue on to actually reaching the world!

10

Reaching the World

Alan Lastufka

Promoting Versus Interacting

Knowing how to properly promote yourself on YouTube is powerful. We spent the past few chapters discussing both productive and harmful ways to interact and promote yourself on the 'Tube. In this chapter, we're going to branch out, travel beyond YouTube, and show you how to promote your videos and channel on other sites, including some permanent solutions should YouTube ever close its doors.

In Chapter 7, I discussed the importance of backlinks. *Backlinks*, in this case, are clickable URL links that, when clicked, take the viewer either to your YouTube channel or directly to one of your videos. You create these backlinks on YouTube by commenting on the majority of the videos you watch, subscribing and *friending* people, joining groups, and doing an assortment of other fun stuff. If you skipped ahead and haven't yet read Chapter 7 yet, I strongly suggest you take a look at that chapter first, because this chapter builds upon the ideas introduced there.

Some of the sites mentioned in this chapter will also make an appearance in Chapter 12 but in a different context and with different uses. This chapter is about promotion. Chapter 12 is about interacting. While the line between the two might be thin or even nonexistent sometimes, I thought it was important to separate them. *Promoting* is a form of sales—you have a channel or a video you want someone to view, and you "pitch to them"; that is, you give them a sales pitch by selling them on clicking through your links and viewing your videos. *Interacting*, on the other hand, is a two-way street and can be much more beneficial, not only for your views but also for the satisfaction you get out of your time spent on the site. Try not to confuse the two, because your friends won't like being pitched to.

Social Bookmarking

We all have a collection of web pages bookmarked in our favorite browser—those pages you visit time and time again but hate typing in to the address bar every time, or they may be pages that have long, complicated addresses. I'll bet you've even sent those links to friends via an instant messaging program or an email message.

Social bookmarking takes this practice and makes it interactive; it makes it fun. Social bookmarking websites allow you to keep your bookmarks online so they are accessible from any computer with an Internet connection. Social bookmarking sites also allow you to easily share those links with your friends and keep an eye on which links your friends (or strangers) are bookmarking.

Some of these sites even allow users to rate your links, pushing good resources or funny videos to the top of the pile while the poorer links remain buried, just sitting in your collection.

Adding your videos and even your YouTube channel link to various social bookmarking websites could potentially drive a lot of viewers to your work.

Examples of popular social bookmarking sites include the following:

- Digg (*digg.com*)
- Del.icio.us (*delicious.com*)
- Furl (*furl.net*)
- Reddit (*reddit.com*)
- Stumbleupon (*stumbleupon.com*)

Digg is probably the most popular but also one of the easiest to manipulate. Users have been known to create what I like to refer to as *Digg clubs*. Basically, stories are featured on Digg's front page, pulling in tons of traffic, based on how many people bookmark, or *digg*, the article. So, if you get enough friends together, you can all agree to digg each others' bookmarks, thus sending all of their, and all of your, bookmarks to the front page. Digg even includes a feature, known as the *shout*, to help you with this. You can shout any article to any single user, or any group of users, via email (Figure 10-1).

Clicking the link from the shout email will take you directly to the bookmark on Digg where you can choose to digg it (Figure 10-2).

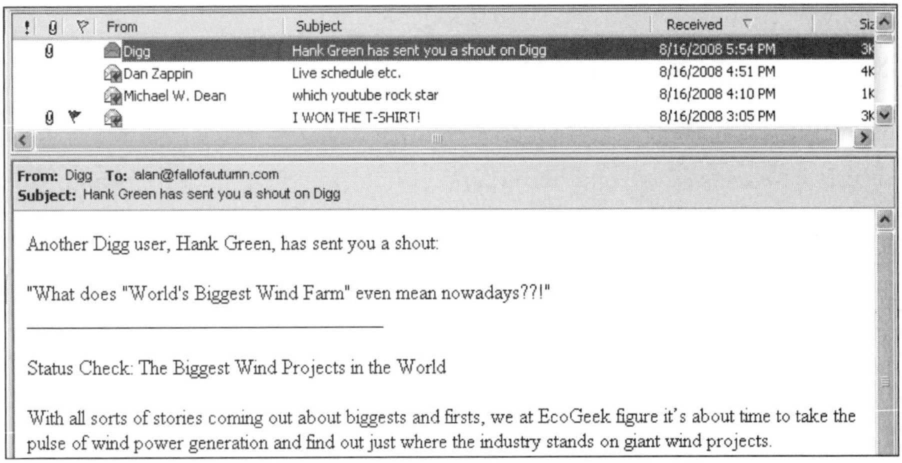

!	⬚	▽	From	Subject	Received ▽	Si⬚
	⬚		Digg	Hank Green has sent you a shout on Digg	8/16/2008 5:54 PM	3k
			Dan Zappin	Live schedule etc.	8/16/2008 4:51 PM	4k
			Michael W. Dean	which youtube rock star	8/16/2008 4:10 PM	1k
	⬚	▽		I WON THE T-SHIRT!	8/16/2008 3:05 PM	3k

From: Digg   **To:** alan@fallofautumn.com
Subject: Hank Green has sent you a shout on Digg

Another Digg user, Hank Green, has sent you a shout:

"What does "World's Biggest Wind Farm" even mean nowadays??!"

———————————————————

Status Check: The Biggest Wind Projects in the World

With all sorts of stories coming out about biggests and firsts, we at EcoGeek figure it's about time to take the pulse of wind power generation and find out just where the industry stands on giant wind projects.

Figure 10-1. Shouts sent from Digg notify you of new bookmarks from your friends.

Figure 10-2. Digging bookmarks on Digg will help push the article onto the front page.

The practice of *digging* for a returned *digg* could be construed to be against the site's terms of use. However, if you truly enjoy your friends' videos, and they yours, I don't see anything wrong with helping to bring as many viewers to that video as possible.

Social Networking

When people hear the phrase *social networking*, the first thought that typically pops into their heads is MySpace. But social networking today reaches far beyond glitter text and Dashboard Confessional profile themes.

A few grown-up social networking sites include Facebook, Ning, and LinkedIn.

- Facebook (*www.facebook.com*) (URL 10.1)

- Ning (*www.ning.com*) (URL 10.2)

- LinkedIn (*www.linkedin.com*) (URL 10.3)

These sites have evolved beyond bored (and boring) middle-school users and have become a powerful way to remain in contact with your friends and even help you make new ones.

I'll discuss more about Facebook and using social networking to help strengthen relationships with viewers in Chapter 12, but here I'll focus on the promotional aspects of social networking. The majority of my friends on Facebook and MySpace are You-Tubers or computer professional freelancers looking for work. My friends on these sites are people working on projects similar to mine and promoting projects similar to mine. They understand, and at times are even interested in, the bulletins I post advertising my latest video, my next live streaming event, or where to preorder this book.

Once you've amassed a few hundred, or a few thousand, friends on these sites, the bulletins and notes you post can become very strong marketing tools. However, I strongly suggest you post only when you have an important project or video and post about each project only once. Too much self-promotion, and eventually your friends will stop clicking through. They'll know before they even read your bulletin that you want them to "click here" or "digg this" or "read more...." Also, make sure you occasionally read and respond to or comment on your friends' bulletins and notes. Not doing so will also make them lose interest, rather quickly, in your postings.

YOUR NETWORKING SAVINGS ACCOUNT

I like to think of networking as a savings account. You can't withdraw anything from a savings account until you put something in. Answer private messages, wish your contacts a happy birthday, and if someone hasn't posted in a while, leave them a comment and ask whether everything is okay. These little deposits could pay big dividends later. Be genuine, though. Banks know when you're handing them fake bills, and your friends will know when you're on a profile-commenting spam spree as well.

If you want to be there, it's better than doing it out of a sense of obligation. Make honest deposits, and when it's time to make a withdrawal, you'll find more than enough help there.

The friends you add on these sites may also open opportunities for you. I was invited to record an unaired clip for the prime-time show *iCaught* on ABC because my friend Jill (*www.youtube.com/xgobobeanx*) (URL 10.4) recommended me and a few others to the producers when they asked for additional guests. I've written guest blog posts for my friend Kevin's (*www.youtube.com/nalts*) (URL 10.5) blog, and he introduced me to a new ad agency, who paid me a few hundred dollars to film a mock ad.

Henry Hartman said, "Success is when preparation meets opportunity." You've prepared. You're reading this book, learning your camera, and mastering your editing program. Now help make yourself open to opportunity by networking with the largest circle of friends that you can.

This may all sound cold and calculating. It does involve calculation, but it doesn't have to be cold. Calculation is simply planning. Promoting and marketing are sales. Sales is all about persuading someone to buy or click or watch, but it can be done in a humane and equitable way. (Remember that sidebar about art karma in Chapter 7?) Again, the flip side to this coin—the interacting and the organic relationships that make all of this worth it—will be discussed in Chapter 12 and are equally, if not more, important.

Blogging

Blogging is a form of online diary, news, or other regularly updated matter. Blogging utilizes blogging software, which started being widely available in 1999. Blogging software is a suite of files installed on a website that enables the automatic posting and organization of single *posts*, or articles, without having to know HTML or web design and without having to upload a new file each time.

Blogs are fun. Most of the social networking sites I just mentioned include blogging capabilities. Blog posts can simply contain text—a quick note about a trip you recently took—or can offer an embedded video or an MP3 download of the music you used in the video.

If you don't want to blog on MySpace or Facebook, numerous dedicated blogging services exist, including the following:

- WordPress (*www.wordpress.com*) (URL 10.6)

- Blogger.com (*www.blogger.com*) (URL 10.7)

- TypePad (*www.typepad.com*) (URL 10.8)

WordPress is the most popular blogging software, especially with YouTubers. Word-Press lets you embed your blog on your own web server space. (Your own server space will cost you a little extra money, but it's worth it; we'll talk about that in the next section.) WordPress will also host your blog for you for free, but it sticks pop-up ads on your site, and it doesn't share that revenue with you. I recommend hosting on your own server. This gives you more control and eliminates the possibility that you'll lose all your work and the web address you've worked so hard to build, should the hosting company go out of business or start charging a lot more for the same services.

Most YouTubers' blogs consist of embedding their YouTube videos in addition to writing text entries and uploading photos about the making of a particular video, a travel journal while attending YouTube gatherings (see Chapter 12), or other special events. You could also include downloadable versions of your videos, pictures, music, or anything else you create inside or outside of YouTube.

Blog posts are also much easier for search engines to index than videos are. Search engines love to read but can't view videos. The words you use in your videos' descriptions and tags help, sure, but blog posts will typically rank higher in search results than videos on the same topic. That said, embedding your YouTube video, and including a short text post with it about the video, is a great way to promote the video via search engines while offering readers a little behind-the-scenes treat in your text post.

Most blogs automatically *ping* search engines, sending out a little bit of data telling them to read a new post as soon as you hit the Publish button. A post on a WordPress blog can show up in a Google search within minutes, whereas it can take weeks for Google (and other search engines) to find a new post on a nonblog website simply by following links.

Michael wrote a great article on blogging software and video blogging. It's called "Put Your Videos on OTHER People's iPods." The article was originally a sidebar in this book but had to be cut for length. You can find it on the O'Reilly Digital Media site:

http://tinyurl.com/6pfuls (URL 10.9)

Your Own Website

Utilizing all these methods for promoting your videos outside of YouTube is important, but as Michael likes to remind me on a weekly basis, any of those sites could be closed, purchased, or remodeled at any time. Websites come and go, and when they go, they usually take all their content with them.

Having your own website, your own domain, can help you safeguard against losing your entire online presence should the websites you rely on fail you. Hosting is cheap, and you can buy a domain name for about $6 a year. You have to also pay for hosting, but that can be as little as $5 a month for a lot of space and throughput. (You can learn more about this, including specific recommendations, in the O'Reilly Digital Media site article about RSS referenced below in the "Blogging" section of this chapter.) Then the space is yours to do with as you wish—blog, upload photos, stream your videos, and embed your videos. Your own website means you are in control of the design, the features, and the content. You can even place AdSense ads on the site and monetize the traffic you drive there (*www.google.com/adsense/*) (URL 10.10). O'Reilly has a good book on using AdSense called *Google Advertising Tools: Cashing in with AdSense* (*http://tinyurl.com/6kjljy*) (URL 10.11).

 Note To create your own website, you need only two things: a hosting plan and a domain name. Usually you can get both from the same service.

When you start your website, your first focus should be content. Many new website owners spend a lot of time on a fancy design only to write "Under Construction" across the top of each page.

Write a few blog posts or upload a few videos before showing anyone your site.

Note Michael adds: And use spell check! Much of the Internet is not spell checked, and people who can spell will often immediately click away from a website full of typos. Even people who cannot spell will notice that **something** is not right, even if they're not sure what it is. Also, read a post out loud to yourself after you write it and before you hit Publish. Blogging software usually includes a spell check feature, but a spell checker won't catch when you meant to type **their** but instead typed **they're**.

Computers are good with math but not yet very good with language.

PRO TIP

When you write a post in Microsoft Word and paste it into your blogging interface, a lot of "junk" is also posted in the form of formatting at the top of the post. This doesn't always show up in the post, but it will show up in things like Google feeds of the post. To delete this formatting, click on the HTML button in your blogging software to show the code, and delete everything above your first word or first image reference.

When you're ready to start driving traffic to your site, do some link exchanges. A *link exchange* is when you place a clickable link (those ever-important backlinks just keep coming up time and time again!) to someone else's website on your site in exchange for them linking to you on theirs.

When exchanging links, you'll want to contact websites similar to yours. First, start with other YouTubers who have their own sites; then, branch out to sites that cover topics related to your hobby—digital video, graphic design, independent music— whatever it is your site might contain.

CONTACTING PEOPLE YOU DON'T KNOW

Michael adds: It's tough to find the right balance on this. People regularly delete link requests from strangers or even block their email addresses as spam. (I know I do, if I get several requests to "check this out!" from someone I don't know, with no other information.) It's tough to find the right balance, especially if you're just starting out and the first email you send to an Internet rock star is you begging for a link. The best links are the ones that people put up for you without you asking. But sometimes people need to be shown what you do so that they will want to link to you. Just put yourself in the frame of mind of the person you're writing and ask yourself whether you're being slimy. Ask yourself whether you can picture your favorite rock star asking someone for a link trade. It might surprise you that many do this, or have done this, but they've probably done it in a way that didn't make people hate them.

It often helps if you write a short paragraph about why the recipient might want to "check this out." That will come across as less random and more personal. You'll probably want to tailor it for the individual person, rather than using a cut and paste. It helps if you say something specific you like about what the recipient does rather than just the generic "I like your videos" or "I love your website."

Even though I'm usually the person getting emails like this, rather than sending them, I still occasionally send them if I come across someone whose work I genuinely dig and think they might dig something specific of mine. And it's not always someone more popular than me. Often it's someone just starting out, but I really think something I did will be of use to them. Live to help poeple; don't just spam people. Be "part of the solution, not part of the problem."

Conversely, don't be sad or mad or write something snitty back if a popular person you don't know ignores your email or even writes back with "Who are you, and why are you sending me this?" People at a star level of media fame often get 300 actionable emails per day.

Once you have some interesting content up and some backlinks, focus on updating regularly. Nothing will keep people returning to your site like new daily or weekly posts. If you update on Fridays, always update on Friday; don't update on Tuesday one week for no reason and then skip a week. Visitors (and more importantly, search engine spiders) love consistent, fresh content.

The exception to this is if you really have nothing new to say on the day you usually update; if you have nothing to say, don't just post something lame. It's a good idea to have a few posts "in the can" that aren't specifically timed to any current event so you can post them if you have writer's block or video idea block one day. In most blogging software, you can write a post and save it as a draft to be posted later.

I also recommend regularly backing up your blog posts. You can easily do this in WordPress using the Manage/Export function on your dashboard. That way if your site is ever hacked or your server crashes, you can restore it from the last backup. (Don't forget to save your media too. The Manage/Export function saves only the text of your posts.)

Signatures

Finally, we're going to make use of forum signatures and email signatures. If you frequent any online forums, you'll have noticed that most of them typically allow you to create a default signature, tag line, quote, link, and/or photo that appears at the bottom of every post you make. Take advantage of these backlinks!

Think about how often you reply to threads on a forum. If you're active on a forum, it would be easy to post a dozen replies to threads on an average day. Now imagine if each of those posts included a link back to your YouTube channel. Other users of the forum who agree, or even those who disagree, with your replies there might be interested enough in what you say and who you are to check out the links in your default signature. All that clicking will lead to additional views and maybe even a few extra subscribers.

Forum signatures, just like email signatures (which I'll discuss in just a moment), are automatically attached to each reply you post. The signature is totally "set it and forget it," meaning you reap the benefits of all those backlinks for very little time or energy put forth setting up your signature.

And, Google spiders love forums. Google ranks its search results by *page rank*, a mysterious algorithm that only those "behind the curtain" inside Google know the exact details of. However, it's been reported that an important factor in page rank is the number and quality of a page's inbound links, especially anchor-text links.

Email signatures work in much the same way, only you don't need to worry about your emails being indexed by the search engines. Figure 10-3 shows my current email signature. The signature is the part below the double line and above the "Original Message" part I was replying to.

Figure 10-3. You can include links in your default email signatures to help promote your websites.

Try to keep your link list manageable. No one will click through to 10 different sites from your email, and even my three might be considered too many by some. And some spam bots perceive any email with more than a few links to be spam. But, all three I use are clearly listed and serve very different purposes. Again, think about the number of emails you send on any given day. Making sure a link to your YouTube channel appears at the end of each one will inevitably lead to more clicks and more views.

Moving Forward

We've discussed getting involved on YouTube. We've gone step-by-step through promoting yourself on YouTube and across numerous other sites. We've shown you how to set up your own "worldwide TV station from your room" with blogging software.

If you've actively followed along so far and if your content is decent, you've worked hard, and kept at it for a while, you should be receiving a few thousand views on each video you upload. You may even be ready for the YouTube Partnership Program, a way to monetize your videos. Read on to Chapter 11 for more information about the Partnership Program, how to apply, and what it will mean for your channel.

11

Money Money Money!

Alan Lastufka

Monetization

Monetization, an old financial term, has recently been applied (perhaps even overapplied) in the Web 2.0 world. In short, when you *monetize* your online content, in this case your videos, you find ways to make money from your free content.

You can monetize your online videos in several ways. You can host them on websites that offer revenue sharing, as YouTube does with its Partner Program (usually referred to by users as the Partnership Program). You can run commercial preroll or postroll ads on your videos (see the "Preroll and Postroll" sidebar). Or you can put in paid product placements. Some of these techniques are easier than others to accomplish, but you will take a look at each technique throughout this chapter, with the focus, of course, being on YouTube.

Before we get into it, though, let me preface this chapter with a word of caution: Don't expect to retire off your online video earnings any time soon. In fact, for the first year, don't even expect to cut back to part-time at your day job. Building an online presence, especially an online audience, takes time. Capitalizing on that audience takes even longer. It can be done, and my current work schedule proves that. I work every day at my "day job" from 6 a.m. to 10 a.m., and then the rest of the day is mine. To write. To edit a new video. To make popcorn, kick back, and watch other people's YouTube videos. To take a nap and then watch reruns of *Monk*. It's up to me because I've put in the legwork and the hours networking, building an online presence, and capitalizing on that work.

The previous chapters offered guidance on the *work* that's required to build your audience and your name. This chapter tells you how to get *paid* for that work.

Keep in mind that when you're self-employed, you have to actually *get some work done*. A lot of work. Many people, when they don't have a boss telling them to work, don't do anything. Being happily self-employed requires discipline. This is important enough that I'll say it twice. Being happily self-employed requires discipline.

PREROLL AND POSTROLL

The terms *preroll* and *postroll* refer to short 10- or 15-second commercial advertisements that run before (pre) or after (post) your video. Some video-sharing websites offer these ads as a way to generate revenue, and some of those sites share that revenue with their content creators. Preroll and postroll ads are generally viewed as the most obtrusive form of advertising in conjunction with user-generated videos. The viewer must wait to watch the content they want to see until the ad ends. It's not as easily ignored as an InVideo ad or an AdSense ad near your video, both of which are discussed later in this chapter.

The sites offering preroll and postroll ads solicit advertisers for you. You simply upload your video, and the ads are there waiting for your viewers. On sites that do not offer these ads, you will have to run the marketing side as well, meaning you will have to contact companies that may be interested in advertising alongside your content and pitch them. This can be more profitable, because you don't have to split the revenue with the hosting site. It is also much more time-consuming because you may spend hours landing a single client. I don't recommend it, at least not until you've mastered doing it with a partner like YouTube.

YouTube currently does not run preroll or postroll ads. (And many users hope they never will.)

What Is a YouTube Partner?

YouTube Partners are users who have applied for, and have been accepted into, the YouTube Partner Program. YouTube Partners receive a portion of ad revenue for their videos. This money is paid by advertisers to YouTube so that the site can run ads on Partner videos. YouTube then splits this money with the Partner. Currently, 45 percent goes to YouTube, and 55 percent goes to you—the YouTube Partner and content creator.

Why doesn't YouTube run ads on everyone's videos to make more money? Well, more than 10 hours of video are uploaded to YouTube every minute, making it nearly impossible for YouTube to screen it all. Some of that content is copyrighted, and if You-Tube were to run an ad next to a video containing someone else's copyrighted material and profit from streaming the copyright owner's works, YouTube could easily be sued by the original copyright owners. By not running ads on infringing material and by agreeing to remove any infringing material at an owner's request, it is much less likely that the original copyright owner would file a lawsuit against YouTube.

But YouTube still needs to make money; after all, it is a business, with employees and electricity bills and office rent to pay. And it has to maintain those mighty banks of server computers powered by genetically modified superhamsters on precision-bearing exercise wheels in that secret underground lair. (Just kidding. There probably aren't any superhamsters.) So by creating the Partner Program, YouTube, for the most part, protects itself from advertising on infringing material and still brings in some revenue to cover its costs. In return for you promising never to upload copyrighted material, you are given a few perks: some additional channel design features, branding options, and, oh yeah, the baby lion's share of the ad revenue split.

Any user can apply to be a YouTube Partner, provided the program is available in your country. But YouTube insists you meet some basic requirements to be a Partner. Read on to see whether your channel might qualify and, if it does, how to apply.

Do You Qualify?

YouTube states three requirements for its Partners:

- You create original videos suitable for online streaming: This means you cannot upload any copyrighted material, even if it falls under fair use laws. You cannot upload any mash-ups (videos that were created using parts of other videos). In addition, while YouTube doesn't have any hard and fast rules about profanity in videos, your videos must be appropriate for companies to want to place their ads next to. Too much violence, vulgarity, or sexuality will scare off potential advertisers (and their revenue).

- You own the copyrights and distribution rights for all audio and video content you upload—no exceptions. This is a point so important that YouTube said it twice. It is possible to get distribution rights or limited distribution rights from copyright holders to use commercial songs in your videos. In my letterpress documentary, *www.youtube.com/watch?v=yE0OoWX6TQs* (URL 11.1), I received written permission from numerous bands, including Fugazi, for use of their music in a commercial setting (I am receiving ad revenue sharing on that particular video). To obtain permission, I simply contacted the bands' record labels. Some labels will be more open to the idea than others, and it can take a lot of work. It's usually a better idea to use music you create yourself, music by friends who give you permission, Creative Commons music that allows for commercial use, or royalty-free music from sites such as PodSafe.

- You regularly upload videos that are viewed by thousands of YouTube users: You probably shouldn't apply for the Partner Program one hour after uploading your first video. YouTube wants to know you have an audience and that you can bring viewers' eyes to your videos. Spend some time with the earlier chapters in this book. Build your audience, slowly and honestly. Then, once your videos routinely receive more than 1,000 views a day, you should apply. I receive an average of 5,000 views per day, whether I post a new video that day or not. That's my back catalog working for me. Viewers find videos via searches, related videos, video responses, and the various other features YouTube has in place.

While it may be frustrating to wait before applying to the Partner Program, even if you were accepted with minimal views, the money being made on those views wouldn't add up very quickly. As partners, it's part of the agreement that we're not allowed to disclose the exact amount we're making, but unless you routinely receive a few thousand views per day, you will be very disappointed with your earnings. Users at my level, around 10,000 subscribers and an average of 150,000 views per month, will make a couple hundred dollars per month. Users at the levels of LisaNova or What-TheBuckShow, however, can make six-figure annual incomes from YouTube. If you've followed the advice Michael and I have given you up to this point and your channel is rockin', then read on, and send in your application. Our fingers are crossed for you.

How to Apply

Applying for the YouTube Partner Program is easy and takes only three steps. You must be 18 or older to apply for yourself. If you're younger than 18, YouTube asks that your parent fills out the form for you. Your application will be tied to a checking account, so you can't fill out the application under your name if you're not of age.

When you are ready to apply, visit the Partner Program portal page at: *www.youtube.com/partners* (URL 11.2), as shown in Figure 11-1.

Scroll to the bottom of the YouTube Partner Program portal page, and click the Apply Now button. After clicking Apply Now, you will be taken to an online application form (Figure 11-2).

Figure 11-1. YouTube Partner Program portal page.

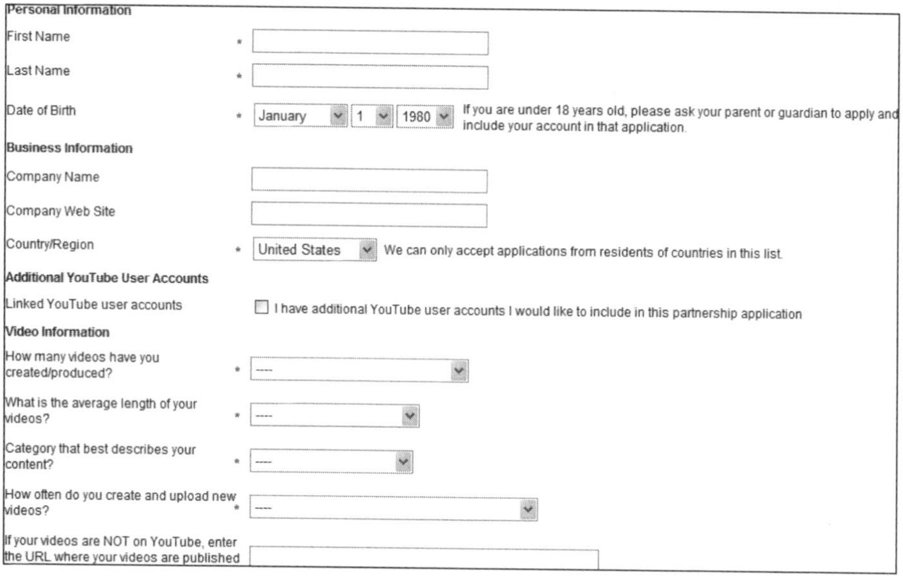

Figure 11-2. Online Partner Program application form.

The Partner Program application form asks you for standard personal information, such as your actual legal name (most Internet users go by screen names, or *handles*), and for information about your videos. The information you provide here helps determine whether you're approved for the Partner Program, so I advise you to remain professional and honest.

After you've filled out the application, click Review Application. You will not be instantly approved. It will take some time (between one and ten days, usually about three days) for YouTube staff to review your application and your videos. Be certain your videos adhere to some of the guidelines I mentioned in the previous "Do You Qualify?" section.

You will be informed via email when YouTube has made its decision. If you are rejected, your email may state a reason for YouTube's decision. Don't become frustrated. You may reapply as often as you'd like. If YouTube offers suggestions that may help you get approved on a future application, follow those suggestions before reapplying. You may be rejected for any number of reasons, but the most common ones I've heard are lack of views, lack of subscribers, and lack of interaction with the YouTube community. You can follow the advice throughout this book to rectify all of these issues. If approved, you'll be sent an email that looks like the one in Figure 11-3.

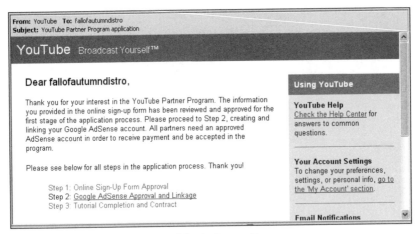

Figure 11-3. YouTube's welcome email for the Partner Program.

The email you receive will provide a clickable link to the next step in the process, creating your Google AdSense account and linking it to your YouTube account. Click the link YouTube provides, and follow the onscreen instructions. If successful, you will receive the second of three emails (Figure 11-4).

The second email gives you a link to a tutorial and contract. The tutorial explains how to submit your videos for revenue sharing, covers what's allowed in uploaded Partner videos, and provides additional YouTube contact iformation, exclusive for Partners.

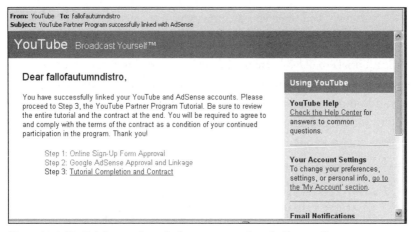

Figure 11-4. YouTube's second email after acceptance into the Partner Program.

You can digitally sign the contract after viewing the tutorial. Read your contract carefully; when you digitally sign it, you're legally agreeing to the terms and conditions stated. After you complete the contract successfully by agreeing to the terms and digitally signing it, you will receive a third and final email from YouTube (Figure 11-5).

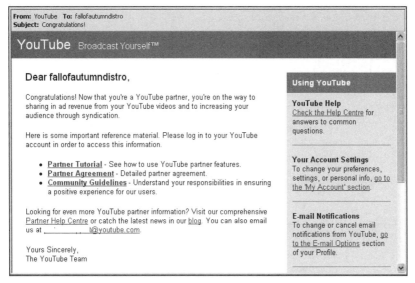

Figure 11-5. The third and final email welcoming you to the Partner Program.

I suggest holding on to these emails (as I have) so that you can refer to the useful links provided should you need them later. As you can see in Figure 11-5, the third email includes links to the tutorial, a copy of your online contract (the agreement), and the exclusive Partners-only email support address. (I've blanked mine out in the figure. It's a secret. Get your own, and once you do, don't give it to anyone.) These are good links to have handy, if and when you need to refer to them.

After this third and final email, your channel will be upgraded to a partnered channel within 48 hours. Ad revenue sharing isn't the only benefit of a Partner channel; you will also receive additional branding options for your channel (see the following section), which allow for image banners and additional channel layout options exclusive to YouTube Partners. (And no ten-minute limit on video length!)

AN INTERVIEW WITH GEORGE STROMPOLOS, HEAD OF THE YOUTUBE PARTNER PROGRAM

Alan Lastufka: What is the YouTube Partner Program?

George Strompolos: The YouTube Partner Program enables original content creators and media companies to participate in ad revenue sharing and syndication options through YouTube. The high-level goal is to help compensate active users for their creativity. It's a democratic, performance-based program where more video views translate to more revenue for a given Partner.

Alan: What kinds of steps are required to become a YouTube Partner?

George: YouTube partnership is a privilege reserved for active uploaders and media companies that make a positive contribution to the YouTube community. A good first step is to create a channel, just as any user would do, and begin building an audience by uploading videos and interacting with other users. Exactly what you upload is up to you, but it's important to remember that we partner only with original content creators. In other words, you must own the copyrights and distribution rights for all audio and video content that you upload. Once your videos are consistently generating thousands of views, simply apply at *www.youtube.com/partners*. (URL 11.3) YouTube will review your application and respond ASAP.

Alan: What will get someone turned down for the Partner Program?

George: Most users get turned down because they simply don't have enough views or subscribers to qualify for partnership. The good news is, users without enough views or subscribers can continue uploading and may apply again at a later date. In addition to this "popularity" qualification, users with a history of violating YouTube's terms of use will not be accepted as Partners. Such violations could include uploading content that you don't own, uploading obscene content, spamming or harassing other users, and attempting to "cheat the system" for more views or subscribers.

Alan: What is YouTube looking for in a YouTube Partner?

George: We look for originality, creativity, popularity, upload frequency, and active community participation. Our Partners range from video bloggers and sketch comedians to major media companies and celebrities. We're constantly amazed by the level of creativity out there!

Alan: What can partners expect from YouTube once enrolled?

George: New Partners will immediately notice several new branding options are now available such as channel page banners, channel page autoplay, watch-page banners, and several other cool bells and whistles. Partners have an option to select which of their videos they want to monetize. YouTube displays ads on monetized video watch pages, and Partners receive the majority of this ad revenue. If a Partner's monetized videos are generating a lot of views, their revenue share can quickly become very substantial. Partners with relatively fewer views also seem to appreciate the added income. YouTube also likes to feature our Partners in several areas across the site, and we routinely connect our advertising Partners with our content Partners for sponsorships and special projects.

Alan: Why was the Partnership Program instated?

George: The high-level goal is to help compensate original content creators for their creativity.

Alan: How did you get your job at YouTube?

George: As a tech enthusiast with a passion for media and entertainment, I've been working in the Silicon Valley since graduating from U.C. Berkeley in 2002 with degrees in Mass Communications and Business. I got started at *Wired* magazine, moved to CNET and then Google, and ultimately found my way over to YouTube when the company was acquired by Google in 2006. As a hobbyist animator and musician myself, it's been extremely fulfilling to help build the YouTube Partner Program, which is truly revolutionizing independent distribution. The best part is, we're just getting started!

Branding Options

In addition to monetizing your videos, YouTube Partners are offered some unique channel design options called *branding options* (Figure 11-6).

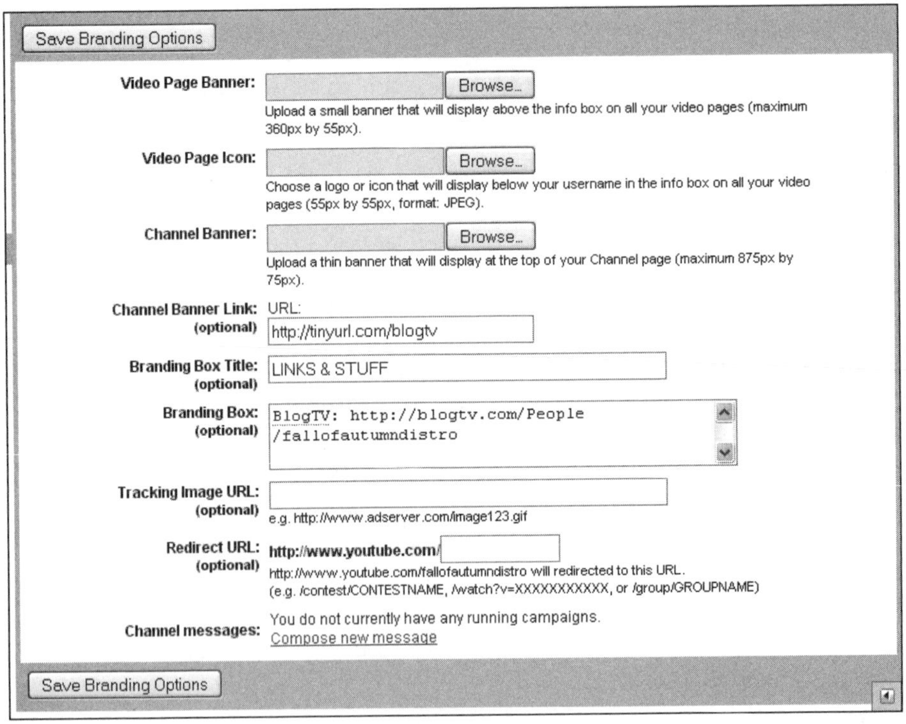

Figure 11-6. YouTube Partners are given exclusive branding options for their channel design.

The branding options include image banners on both your video pages and your channel page. These banners are clickable, and you can choose which link your channel banner takes the viewer to when it's clicked. This is a great way to advertise your personal website or products you might have for sale (T-shirts come to mind).

The branding options also include a text box on your channel page, known as the *branding box*. For non-Partners, the channel description allows for only one clickable link. This additional text box for partners allows for unlimited clickable links to be added to your channel page. You can't, however, place any other HTML code within this box, meaning no additional images or widgets; you can use only *http://* links. Find these additional branding options under your Account link; they won't be visible to non-Partner channels.

> **Note** You can get customized T-shirts (and a lot of other cool *swag*, that is, merchandise) made with no setup fee from *www.cafepress.com* (URL 11.4) or *www.lulu.com* (URL 11.5). These two sites allow you to upload your own art with words and pictures and set up a linkable online store. They take care of the printing and shipping and send you a cut (determined by you—you set the markup from the wholesale price).
>
> Once you start selling a lot of items every week, you can get better deals by going directly to a T-shirt printing company, paying up front, and mailing the orders out yourself from a P.O. box (don't use your home address for your business; in fact, never post your home address online). But CafePress and LuLu are great for dipping your foot into the *merch* (merchandise) pool without spending any money.

AdSense

YouTube is owned by Google and therefore uses Google's advertising program, AdSense, to monetize its Partners' videos. AdSense (*www.adsense.com*) (URL 11.6) is a program open to anyone who creates an AdSense account. Outside of YouTube, AdSense is used by webmasters and bloggers to display text and picture ads alongside their content, in hopes of making money from these ads. Google provides the ads and splits the revenue with these website owners, just as YouTube does with its content creators.

Most AdSense ads are paid on a per-click basis, meaning the webmaster and Google make money only when a visitor clicks the ad. However, some ads are paid for on a per-impression basis, meaning the webmaster and Google make a set fraction-of-a-dollar amount every time a visitor views that ad, that is, visits the page on which the ad is displaying.

YouTube ads are all paid for on a per-impression basis. Ad rates seem to vary from campaign to campaign, because earnings per view vary each and every month. AdSense ads display next to videos uploaded by Partners. You'll need to keep your AdSense account in good standing to remain in the Partner Program. This means you should not try to fraud the system by autorefreshing your videos. You should also not click over and over on your own ads; this gives the impression to advertisers that your videos are more popular than they actually are and breaks the contract you sign with YouTube when you become a Partner. (Both YouTube and AdSense have really smart software to detect all fraud techniques, and you will get caught.)

PRO TIP

Advertisers and YouTube attempt to target relevant user-generated content with their ad campaigns. If a cell phone company wants to advertise with YouTube, for example, YouTube will attempt to display those ads on videos whose descriptions and tags include words like *wireless*, *cell*, *phone*, *voicemail*, and other keywords relevant to cell phone companies.

You can use this knowledge to target specific advertisers, although you will end up with a quandary. Should you keep making the videos you want to make, or now that you're a Partner, should you try to tailor your videos specifically to try to make more money?

In my experience, and in that of a lot of people I know, the best mix is about 90 percent of the first and 10 percent of the second. That is, keep making videos the way you did that originally attracted your audience, but keep it in the back of your mind that you're making money on them now. Don't try to "sell out" and make videos just to make money. Keep making the content from your heart that brought people in to begin with, but don't discount the idea of doing something here and there to better monetize your work, as long as that doesn't overwhelm your videos and turn them into something else.

First, people will unsubscribe if they feel they're being overtly "marketed to." YouTube is an alternative to TV. If you make your channel too much like TV, people will go look at another channel.

Second, you're not going to make tons of money, just some money, so you may as well still have fun doing it, rather than making video production an unpleasant day job. There's no point in working toward quitting your day job if you simply replace it with another job that doesn't make you happy (and doesn't offer health insurance!).

There is an exception to this "don't make targeted videos" suggestion, but it's only for mega-YouTube stars, like the top 500 or so channels on YouTube. It's covered in the next section, "Sponsored Videos."

While the majority of AdSense ads display next to the video (Figure 11-7), a second form of advertising on YouTube, known as an InVideo ad, displays in the video. InVideo ads are worth much more than the ads that display only next to your videos. InVideo ads display within the bottom 20 percent of the video window, for 15 seconds, starting at the 15-second mark in the video (Figure 11-8). These ads may be a little more intrusive than the AdSense ads, but the user can close them immediately; and, as I said, the InVideo ads pay much better than the ads that display only next to your videos.

Figure 11-7. AdSense ads display next to videos uploaded by Partners.

Figure 11-8. InVideo ads display in the lower 20 percent of the video window and pay more than the ads that display only next to your videos.

InVideo ads pay better, which means they cost advertisers more. These ads are not commonplace. InVideo ads are typically reserved for bigger campaigns. If you see these ad campaigns while watching others' videos, you can amend your *relevant* videos' descriptions and tags (see the "PRO TIP" sidebar) to target that particular ad campaign.

With any luck, your amended video will start displaying the current InVideo ad campaign.

Sponsored Videos

If you're really in the 'Tube game to make a living, creating targeted, sponsored videos will bring in a substantially higher salary than ad revenue sharing alone. Some content creators have built very large audiences and can offer advertisers guaranteed view counts. Think of it as a celebrity endorsement for a product, but the celebrity is you. Numerous YouTubers have created sponsored or targeted videos in the past. You have to have tens of thousands of views on each of your videos to do this.

Kevin Nalty, *www.youtube.com/nalts* (URL 11.7), has created numerous sponsored videos for products such as Mentos *www.youtube.com/watch?v=Eu21hMjLPCA* (URL 11.8) and Holiday Inn, *www.youtube.com/watch?v=V970qvl14zk* (URL 11.9).

When creating sponsored videos, you must do all the legwork of contacting, pitching, and securing an advertiser. (If you end up in the top 50 or so on YouTube, however, companies will contact you.) You should have a number of video ideas to pitch (remember the "elevator pitch" from Chapter 2?) because some ideas will resonate better with corporations than others. Sponsored videos can be worth an average of $3,000 to $10,000 per video. Make sure you figure all of your costs into your price, including any props used in the video and the time you will spend editing, uploading, and replying to comments. Although replying to comments is a lot of fun and helps build a relationship with your viewers, it is also very time-consuming. There are rumors that some A-list 'Tubers hire people to do commenting for them, posing as the channel owner. This is not currently a violation of the terms of use, but that could always change.

Don't think of a sponsored video as a typical commercial. You should write, direct, and edit your sponsored video like any other video you're making for your channel; just remember to include the product in there somewhere. Funny videos less than 2 minutes in length will do better than any other format. nalts' Mentos video is one of the better examples of a sponsored video on YouTube. The product is the star, but the video doesn't feel like a commercial; it feels like any other prank or comedic video user nalts would make anyway (see Figure 11-9).

Figure 11-9. YouTube Partner nalts' attempts to sneak an oversized roll of Mentos into the movie theater in a video sponsored by Mentos.

Selling Out, Again

Occasionally, non-Partners criticize partnered channels for "selling out." I know we've mentioned this previously in the chapters about promoting yourself, but unlike before, where your gain wasn't concrete, here you are receiving cash in hand, from huge corporations, for your creative work. You need to decide before applying whether you will let advertisers indirectly censor your creativity. As a Partner, you must tame the language, violence, and sexuality of your videos to fit within the Partner guidelines, which are typically what national corporations would deem acceptable.

For me, this wasn't a big deal; I rarely curse more than once or twice in my videos (or in my daily life), and the most risqué I've gotten sexually was showing an unused condom onscreen. While I don't have advertisers lining up outside my bedroom door to advertise on that video, it wasn't anything you wouldn't typically see on late-night network television programming. (I've even seen a *used* condom being put into an evidence bag on *CSI*, which airs during prime time.)

Other users might feel restrained by the guidelines of the Partner Program.

Non-Partners have a skewed idea of how close Partners work with advertisers and YouTube. As a Partner, you are almost never contacted directly by any advertiser; that's reserved for the top 5 percent of the top 5 percent of channels. Also, there is no golden key to the YouTube staff attached to your contract. Everything is automated.

When a Partner uploads a new video, they simply choose to include an ad, or to not include an ad, on that video. They do not choose which ad it will be. They do not choose the rates of that ad. The advertisers and their rates are solicited and set by YouTube.

 You will receive some "hater" comments and email as a partner that you would not receive otherwise, but it's all from jealous people who are not good enough to get popular enough to be Partners themselves. Ignore these weenies. Block them.

The Partner Program can be a great way to earn some additional income, or it can even become your main source of income, depending on your popularity. I personally know numerous YouTubers who are living off their YouTube income alone, and some can even afford to hire editors and additional writers with their YouTube income. You can be in their position too, with a good bit of hard work, time, and creativity.

No one got there overnight, though. No one did it with mediocre videos, and no one did it without devoting a lot of time, over a period of time. We'll give you some tips for time management and more in Chapter 12, "Beyond the 'Tube."

12

Beyond the 'Tube

Alan Lastufka

Stepping Out

So, you've established your presence on YouTube. You've made a few casual friends—people who you watch and people who watch you. Maybe you're even a YouTube Partner now and making some extra cash each month from your videos. Is this it? What's next?

Well, next comes building real connections—personal, emotional, business, and creative. The relationships you seed on YouTube can soon grow far beyond the 'Tube's digital walls.

Staying on one site all the time can be a little like walking around with blinders on. Many YouTubers connect outside of YouTube, whether on other websites better geared toward socializing or at YouTube Gatherings.

YouTube Gatherings are public events held across the globe for YouTubers to hang out together, buy each other a beer, or just enjoy an afternoon of Frisbee in the park.

This chapter will point you toward some of the sites I and other YouTubers use to stay in touch and to better interact with each other. You will also learn how to find and attend future YouTube Gatherings. Later, Michael Dean will talk a bit about time management to ensure you don't spend your entire day in front of a computer screen.

Facebook

Originally, Facebook was available only to college students with *.edu* email addresses. The site is now open to the public and is quickly becoming a favorite over MySpace for young adults who want to stay in touch with friends. If you want, you can add me at *www.facebook.com/people/Alan_Lastufka/625314358* (URL 12.1). Facebook offers a solid private messaging system that allows you to email with your friends from any computer. You can add friends and receive updates via Facebook when your friend has a birthday, hosts an event, or publicly posts an update about what's going on in their lives.

If you want to use Facebook as a promotional tool, you can also post links to your videos, your personal site, and announcements about events like upcoming live shows.

Most people feel more comfortable sharing themselves on sites like Facebook because, unlike YouTube, their Facebook updates are private and available only to the friends they've added. Anyone can subscribe to you on YouTube or watch your videos without even having an account on YouTube, but for lurkers to view your posts on sites like Facebook, they must request that you add them as a friend, and you must approve that friend request.

Facebook is simple enough to join and use, but if you get stuck, O'Reilly has a wonderful step-by-step book all about Facebook entitled *Facebook: The Missing Manual.* Check it out here:

http://tinyurl.com/5rt86r (URL 12.2)

Additional similar sites are *www.myspace.com* (URL 12.3), *www.xanga.com* (URL 12.4), *www.linkedin.com* (URL 12.5), and *http://programmermeetdesigner.com* (URL 12.6). MySpace seems to have been taken over by teenyboppers who consider the pinnacle of social interaction to be sending grade-school "Do you love me?" surveys to each other. Xanga seems to me to be sort of a "not sure what it wants to be, so it's got a little bit of everything but not much focus" site. LinkedIn is a good site if you're a freelance worker of any kind and want to network with potential employers or collaborators. ProgrammerMeet Designer is a networking site specifically for computer programmers.

BlogTV

BlogTV is quickly becoming the de facto live streaming video site for YouTubers. Subscribe to my live show at *www.blogtv.com/people/fallofautumndistro* (URL 12.7). YouTube can play back only prerecorded, uploaded videos. YouTube cannot currently broadcast live video. Live streaming video can be a much better platform for interacting with friends and subscribers. Until YouTube offers a live streaming video feature, we must rely on third-party sites like BlogTV and Stickam (we will talk about Stickam in the next section).

BlogTV visitors can watch any of its hundreds of live streaming shows for free. Each show has a host, the person who controls the show and appears onscreen. The host may entertain viewers with a story, a musical performance, or any other type of broadcast (though sex, nudity, and copyrighted material are not allowed). The host may also interact with viewers by reading and answering questions from the show's text chat (Figure 12-1). BlogTV also allows for an additional cohost for conducting interviews or simply hanging out and talking with a friend on cam.

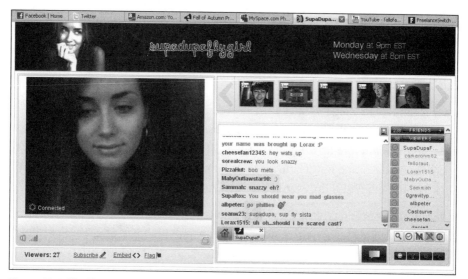

Figure 12-1. BlogTV user SupaDupaFlyGirl broadcasting live.

BlogTV is unique and offers a recording function for live shows, something very few live video streaming services currently offer.

Viewers may access an archive of recordings of any previously broadcast show as long as the host decided to record it. In addition to the video, BlogTV also saves the text to uniquely re-create the experience of being at the live show. BlogTV maintains numerous lists of the Most Viewed videos for the day, week, month, and all-time, making it easy for new users to find popular shows.

Viewers may choose to subscribe to shows they enjoy, with the option of receiving an email every time that show goes on the air. With the random schedules of some broadcasters, this is a great feature to ensure you don't miss any live shows. Viewers may also add friends, a feature that allows users to text message each other in real time, even if they aren't both watching the same show.

I've broadcast a live show on BlogTV for three months as of this writing, and I've found it to be one of the most stable of the streaming video sites. I've also received numerous prompt and helpful replies from BlogTV staff every time I've sent them an email. I highly recommend creating a show on BlogTV and using it to keep in touch with friends, broadcast live vlogs, or host special events outside of YouTube.

You can even download recordings of your live shows on BlogTV and later upload them to YouTube.

Additional similar sites are *www.ustream.tv* (URL 12.8), *www.nowlive.com* (URL 12.9), *www.stickam.com* (URL 12.10) (see the next section).

Stickam

Stickam is similar to BlogTV. Stickam is a free live streaming video service that is also used by numerous YouTubers (you can subscribe to my Stickam shows at *www. stickam.com/fallofautumndistro*) (URL 12.11). Stickam came before BlogTV but has been slow on updates and new features. Most YouTubers started out broadcasting live on Stickam, and many still do, even if they also host shows on BlogTV.

Stickam differs from BlogTV in a few ways. Stickam allows for both public and private live shows, while BlogTV currently allows only public shows. Stickam also allows for up to seven people to broadcast in the same room (BlogTV allows only two broadcasters, a host, and a cohost per room).

Numerous YouTubers have broadcast live from various YouTube Gatherings using Stickam and BlogTV (you'll find more information about YouTube Gatherings later in this chapter). In fact, the very first Stickam broadcast I attended was a live feed from YouTube Gathering 777 in New York, one of the largest Gatherings to date.

Additional similar sites are *www.ustream.tv* (URL 12.12), *www.nowlive.com* (URL 12.13), and *www.blogtv.com* (URL 12.14) (see earlier in the chapter).

Twitter

Twitter is a micro-blogging site. You can follow me at *www.twitter.com/AlanDistro*, (URL 12.15). Twitter allows you to send text updates to your friends. These updates are short, limited to 140 characters by Twitter. You can post updates throughout the day or reply to others' updates.

While most updates are personal posts about who's waiting in line at which credit union and who's having what for dinner (I find this a big waste of time, though a lot of people seem to enjoy doing it), you can also use Twitter to promote your new videos, your live shows, or any other project with a URL attached to it. Most of the people I follow, and who follow me, on Twitter have found a good balance of promotional and personal posts. No one enjoys being spammed, but if you are friends with most of the people you follow, you won't mind being reminded that their live show is starting in an hour or that their new video is up.

Additional similar sites are *www.plurk.com* (URL 12.16) and *www.pownce.com* (URL 12.17).

YouTube Gatherings

YouTube Gatherings are social events held in various cities across the globe. The majority of these events are hosted by one or more YouTubers, though occasionally YouTube staff members attend or assist with the planning.

The first YouTube Gathering, AsOne, was hosted by YouTube user smpfilms in Los Angeles. smpfilms went on to host and cohost other YouTube Gatherings. The 777 event was held on July 7, 2007, in New York; 888 was held on August 8, 2008, in Toronto, Canada; and other Gatherings have been held in England and Australia.

Gatherings are typically well promoted within the community. If you'd like to attend a Gathering, simply keep your eyes open for videos about upcoming events. Most Gatherings do not require you to RSVP, but some will. Make sure to email whoever is hosting the event for details. Also, be sure to plan your lodging in advance; hotels can book up quickly in the surrounding area.

Gatherings are a great way to make new friends or to strengthen existing friendships. Gatherings break down the various virtual walls YouTube constructs; suddenly subscriber and view counts don't matter when you're all sitting down to eat lunch together or walking around happily lost with three or four other Tubers who wandered away from the event (Figure 12-2).

Figure 12-2. Gatherings are littered with mini-DV cameras. Everyone is trying to capture that one moment that will make a great YouTube video or get their picture taken with their favorite 'Tubers.

It's best not to do anything you would regret at any after-party, because there will always be a camera around to catch it.

Between your friends on Facebook, hosting a live show on BlogTV, and following folks around on Twitter, you may quickly and unintentionally spend more time with your online friends than you do with family or than you have for yourself. Throw in attending a Gathering or two, and you will find yourself with no free time at all. Balance is important. Having a life outside of YouTube ensures that you will have something to vlog about while on YouTube.

In the remainder of this chapter, Michael Dean will cover some unique time management techniques.

Time Budgeting

Michael W. Dean

Part of fitting into the YouTube community, or any community, is making sure you have enough time to be part of that community while still reaching all your goals elsewhere and also still having a life outside the Internet.

Having enough time to do everything you need to do online while still maintaining enough of a life offline to be happy is easy. Simply don't sleep much. Well, there other ways, but they involve a lot of budgeting your time...managing your time...basically, *time management.*

At first, you may have to actually schedule for this, but eventually if you strategically try for a mix of all the activities you'd like to pursue and find what works, you can have an online life, an offline life, a job, a romance, a hobby, and even more than five hours of sleep a night. It's hard, but it's doable. I'm doing it.

I'm pretty much an expert at managing my own time, so I'll tell you what works for me. Everyone is different, but some of it will probably work for you.

Flexibility

A key element of time management is flexibility—the ability to get anything done *from anywhere*. I wrote my book *$30 Writing School* while riding the train around Germany and France with my Eurail pass touring to lecture and show my film *D.I.Y. or DIE: How to Survive as an Independent Artist.* Most people would have looked at that as a vacation, but I look at all vacations as working vacations. In fact, while I've traveled more than most people I know; I rarely travel unless it's work-related.

This is not a deficit, and I don't consider it workaholism. I look at it as maximizing my potential and my time remaining on the planet Earth. It is also a function of the fact that I love the work I do and have structured my life to work toward succeeding in the goal of being able to do what I love for a living.

You need to be able to work from anywhere, any time. In fact, I'm writing this on my laptop in a doctor's office waiting room. I bring a pen with me everywhere I go and constantly make notes. The muse, the spark that causes creation, does not make appointments. The muse strikes anywhere and everywhere, and I have to be able to work from anywhere, any time. I often get up in the middle of the night with an idea, scribble down a line or two of notes, and go back to bed. When I don't have paper, I scribble on the back of my hand.

Editing

You also need to be able to edit on the fly. I take printouts of drafts of outlines for books, scripts, and treatments for videos with me whenever I leave the house. The advantages of printed pages are that they require no boot-up time, no Internet access, and no power. Printed pages are great for maximizing a few minutes of otherwise wasted time. Whereas I won't pull out my laptop unless I have at least 10 minutes (2 minutes to boot it up, 7 minutes to open a document and work, and 1 minute to save and shut down), I'll pull out a page and edit it if I have just 2 minutes. I edit in the supermarket line, in elevators, and in line at the post office; when I used to take the bus, I'd edit while waiting for the bus and edit on the bus.

It really helps me edit if I have something to shut out the chatter of world. The whole world community is not something in my head, especially in the form of random strangers bugging me when I'm in public places.

I often bring little foam earplugs in my pocket wherever I go. Or I'll use my iPod with some nonintrusive background music. And when I do that, I don't use little ear buds. I use big headphones (Sony MDR-7506). They cost about 100 bucks from: *www.musiciansfriend.com* (URL 12.18).

These headphones sound great and cut out about 90 percent of the ambient sound. They sound better than ear buds, shut out more, and keep people from trying to talk to you. Nothing says "Leave me alone; I'm working" like big padded headphones.

Steady Work Beats Frantic Work

Join a gym and any personal trainer will tell you this: Working out three days a week for the rest of your life will get you in much better shape than working out six days a week for six months, getting burned out on it, and giving up. The same is true with artistic work. It's better to work steadily on things and make a habit of it until it's second nature than to work all day and night on something for a few days and then give up.

I always have five to seven projects going at one time, usually one long-term project, several shorter-term projects, and a couple of one-day projects. Managing them takes concentration. I have ADD (*http://en.wikipedia.org/wiki/Attention-Deficit_Disorder*) (URL 12.19), but through careful planning, note taking, practice, and the power of computers, I've turned ADD from a deficit into a plus. (Also, I developed the Dean One-Page Plan, which helps immensely. I'll talk more about that in a minute.)

Multitasking

Slipping in and out of one thing and into another is easier for some people than others. I find that my ADD actually makes it easy to do this if I keep my mind relatively clear, don't hold on to resentments that take up head space, plan my day/week/month and year in advance, and update regularly.

Be careful. Multitasking is good if you're on a computer. It's not good if you're driving. (Incidentally, California just passed a law that makes it illegal to talk on a cell phone without a headset while driving, and I think it's a good thing.) Multitasking on the computer for me involves having five or six windows open at the same time in several programs and going back and forth from one to the other when I get bored or as emails come in. It works. It might seem overwhelming to you, but with practice and care, it enables me to do more in a given time period than if I'm doing one thing at a time.

Occasionally, I mess up and send the wrong email to the wrong person or cut and paste the wrong signature line into the wrong email, but the worst that can happen out of that is not a car crash but, rather, me irking or confusing someone.

Be on a Mission!

As I said, I wrote this entire section while waiting to see the doctor (back pain, from working on this book). In that time, everyone around me read a few magazine articles and watched some TV. I wrote something meaningful that will be read by probably at least 100,000 people. Part of the reason I'm able to do that is because, overall, I believe entirely in everything I do. Sure, occasionally I have little doubts, but they do not stop me. I wake up every morning absolutely sure, beyond any doubt, that today I will change the world. And I do.

This is not in a melpy (self-pity), "I think I'm good; I love you, Michael" affirmation way (which is well parodied by LisaNova in her very funny *Affirmation Girl* videos here:

www.youtube.com/watch?v=e19NrkUcFEQ (URL 12.20)

and here:

www.youtube.com/watch?v=h1EIB1MNV8k (URL 12.21)

No, I don't give myself affirmations in the mirror each morning. I don't *need* to. I am absolutely sure I am absolutely on the right path. And that certainly makes time management easier. If you know you have something valuable to contribute to the world, you can create this conviction in yourself. I've had it since I was a little kid, before I really had anything to say. But even then I knew I would do great things and had the fortitude and internal moral compass to pull it off.

> **Note** If you want to do everything for evil, rather than good, do **not** foster an "I am on a mission, and I will change the world" mentality in art. Go into advertising or marketing instead. Sell soap and widgets and cars and clothes to the washed masses. At least then you'll be relatively harmless.
>
> Remember, both Adolf Hitler and Charlie Manson were frustrated artists.

Effective Weekly Planning

Several popular time management systems have been marketed to consumers. I don't use any of them. They might work for you, but they seem far too much like school, or something I would have to do at an office job, to me.

Some of these systems are sort of straightforward, like the 43 folders idea:

http://wiki.43folders.com/index.php/Tickler_file (URL 12.22)

You set up 43 physical folders (one for each day of the current month and one for each month of the year); put notes, plans, contracts, and such into each; and then check back regularly. But even then, there's a whole philosophy to the methodology of attacking tasks and goals, using a flowchart, changing your whole life around, and using a logic map to deal with things. And you're supposed to buy books to tell you how to do this.

Some of these systems, like the Franklin Covey Day Planner (*www.franklincovey.com*) (URL 12.23), seem almost cult-like to me. People I've met who are into a system like this are really into it, and some of them have a religious fervor about it. This system not only involves buying a special calendar book and carrying it with you everywhere, but it also has a lot of books, software, carrying cases, and even weekend seminars associated with the system. It can cost a lot of money, and they have a chain of stores devoted to this one system. In some companies, you practically have to run your life with the system, at work and beyond, in order to work there.

As you can see, a lot of time management systems exist. Some cost money, some are complicated, and most divide the world into two types of people: "those who use our great system and the losers who don't."

What many of these systems have in common is this: They involve a lot of associated books, coaching, software, and other things that are making a lot of money for someone. That doesn't seem like time management to me; it seems like time manglement: It mangles your time in order to try to save you time.

Thus, I would like to demonstrate to you my free, and much simpler, alternative. It's a time management system that makes me able to write many books, produce and direct videos, upload YouTube videos, make a living at home, remember to buy cat food, and still have time for a social life and even get some sleep. I call it the *Dean One-Page Plan*.

The Dean One-Page Plan

It's simple. It's free. And you can start doing it right now.

I organize my life with one one-page document per week (Figure 12-3). Every Thursday (you can start on any day you like), I print a page from a simple three-column, many-row template I've created. The three columns are labeled "DO," "BUY," and "CALL."

DO	BUY	CALL
Teach DJ MORE web admin and book admin. Inc. Lightning Source.	Charger for iPod	Michele, re: layout of images in YouTube book
Work on chapter 16 of tube book	iPod for wife	~~e-mail Woz with Interview~~ Interview request
~~Register chipped kitties~~	~~CD cases~~	~~Follow up on SPCA~~
CTH on BitTorrent	shotgun shells	Dad's birthday is Thursday
~~Edit PODIO book~~	~~Book markers~~	
Record more for other podiobook	New coffee machine	George EARth
Deposit book check	~~Pay PO box rental~~	
Order more copies of novel, send out orders	Get will notarized	Jez
vacuum & Back up		Lydia
	Upload new YouTube video ("50 things")	London

To blog:>
- Pod gear
- Euro trip 2005 pix
- "Home is where the rights are"
- Acre = 43,560 sq feet. Park by house is 7 acres
- "Simple Pleasures" eBook
- Tom Waits talks about Selby, Tom Waits / Nick Cave story

Figure 12-3. The Dean One-Page Plan in action.

Here's how it works:

- Things in the "DO" column are things I need to do in the coming week.

- Things in the "BUY" column are things I need to buy in the coming week.

- People in the "CALL" column are people I need to contact in the coming week.

"DO," "BUY," and "CALL" seem to cover almost everything I need to think about in a given week, month, or year.

I include email and "mail something to them" in the "CALL" column. You can change it to "Contact" instead of "CALL" if that makes more sense to you, but I like the one-syllable-each simplicity of the sound of "DO," "BUY," and "CALL."

The bottom of the page has a section (in my example, it's called "To blog") for notes of stuff I might not get done this week that I want to carry on to the next week.

I keep the template for this document on my desktop computer. I print it once a week (twice a week on a week with many small tasks) and scratch out items as I complete them. And I write things with a pen in the printed copy as they arise. At the end of a week, I print the list for the next week, removing items that have been done and adding items to do. It really works well in my life.

I've done a blog post about this; you can download my template for it and also see a better close-up of the scanned image shown earlier here: *www.stinkfight.com/2008/07/02/the-dean-one-page-plan/* (URL 12.24).

I deal with long-term goals by getting a calendar every December. I usually get one with kitty cats on it, but you can get one with whatever you feel like looking at for a year. Make sure you get one with enough space to write a few things for each day that you need to (like an inch-and-a-half square). If you get as busy as you'd probably like to be, you'll need it. I also write notes in the margin between the calendar itself and the photo at the top for tasks I need to do at some point this month (Figure 12-4).

The Dean One-Page Plan could change your life and is worth more than the price of this book on its own.

Figure 12-4. The Dean One-Page Plan auxiliary calendar system.

Multipurposing

I take a lot of care with writing emails. Sure, I do them lightning fast, but I think lightning fast. And I spell check emails and often read them to myself in my head to see whether they need any tweaking before I hit Send. I am careful with all emails, whether it's trying to get a well-paid job that will finance my operations for a year or whether it's a simple "thank you" to a fan. In fact, I'm so careful writing emails that they're often so good that I want to use them elsewhere.

I feel it's my right to use my side of any email I write, because it's mine. I never publish the other person's comments without permission, and I never use anything of mine that is specific to any one person, but if I write something to someone else, it's mine to reuse, without needing to ask that person. By doing this I can often "kill two birds with one stone" in a way that is *very* effective time management.

A good example of a place I did this is in the "Ethical Hacking" sidebar in Chapter 9 that I wrote. It was lifted pretty much word for word from an email that I sent to a friend, when we were just riffin' back and forth about anything and everything.

Learn to Type. Fast.

It goes without saying, but typing is basically the main way that information gets put into the Internet and into the world these days. Sure, a YouTube video is not typing, but planning it involves typing, and answering your email is, writing promos is, posting comments is. And dealing with most problems as they arise is mostly done via typing these days. I took a typing class when I was 14. Back then, it was on typewriters. I was the only guy in the class. Everyone else in that class was doing it because they wanted to be a secretary. I did it because I wanted to be a writer, wanted to change the world, and knew I couldn't do that hunting and pecking.

I type 110 words a minute. With great accuracy. And I am really quick with using a mouse and doing keyboard shortcuts to open, save, edit, and close documents. I'll spit out 1,000 words before breakfast—seriously—and they're usually pretty darned good.

You don't need to type 110 words a minute to change the world, but you probably need to be able to type at least 40 or 50 words a minute to keep up with the entry-level amount of work you're going to need to do in order to really make a mark on YouTube, or anywhere.

Internet Addiction

I've heard a lot of talk about Internet addiction, basically starting about a year after the Web existed.

My laptop is turned on about 18 hours a day (from the moment I wake up until the moment I go to sleep), although I'm not on it virtually *all* that time. I'm on it a lot of that time, though—probably 362 days a year. I actually have another computer on in another room that runs 24/7, but it's acting as a server, putting info out into the world, and it's always daytime somewhere in the world. I might be considered an Internet addict by some, because if I get up in the middle of the night and go to the bathroom, I'll often go into that room and check my email before I go back to sleep, usually only if I'm expecting a crucial work-related email. I do business with people in Europe often, so their emails often come in the middle of the night and sometimes require a quick response.

> **Note** You don't have to answer every email immediately. I usually answer all emails ***promptly***. That doesn't mean this minute; it usually means within a day. If it's from someone I'm doing continual exchanges with, like Alan and I writing this book, I often answer almost immediately, because it's more of a conversation than a letter.
>
> I've created an email folder called "Answer Later," and I put emails in this category into the folder; I go through the folder once a week or so and see which one is "ripe" enough to answer. To be honest, I do this most often with fan mail, especially if it's very long and overly glowing and the person wants to ***know*** me and I don't think I'll want to know them.

I've been accused of having Internet addiction. But I don't think I do. (Then again, many drug addicts don't believe they're drug addicts.) The reason I don't think I am an Internet addict is because it doesn't interfere with other aspects of my life and because I use the Internet to *get things done.*

This may be a personal bias, but I think gaming online 18 hours a day is a waste of time. For that matter, I think being on YouTube 18 hours a day might be a waste of time, because you don't have a lot to show for it in the long run. With the games, I guess you do produce something. You produce a high score if you're good and gain the respect of other gamers. And on YouTube, you make people laugh or think, and maybe make a little money, but it's entirely dependent on a system that could disappear at any time. Any day of your YouTube life you could turn on the computer in the morning and find that the site has been sold and is a different animal now. Or you could find it completely gone, forever.

That's one reason I encourage people, if they're making timeless videos—ones that will be interesting past next week—to put their videos on an RSS feed so people can download them and watch them on portable devices. (You can learn more about that in Chapter 10.) I encourage people to use their experience on YouTube (or gaming or whatever it is you spend way too much time doing online) to think about it and expand it into other forms of media or "metamedia" (media about media) that lives on even when the machine is turned off.

Hell, even though I love computers, I may be old fashioned, in a way. I like books (like the one you're holding in your hands). Write a book if you can. Use YouTube (and the rest of the Internet, particularly Wikipedia) as a self-administered university to get smart and skilled and knowledgeable enough to write some books. (I recommend my book *$30 Writing School* if you want to know how to actually get one published.)

Books are very archaic technology. They're stories smashed into ground-up pieces of dead trees. But books are one of the few common art forms these days that would exist if the Internet—nay, the entire power grid—went black. If civilization failed, you could still read a book by the light of the sun (or even by a full moon) and get as much out of it as you could in a fully functioning technologically advanced society. Of course, a YouTube book wouldn't be of much use then, but if nothing else, the Dean One-Page Plan could help you rebuild society, and that's in this book. You don't need a computer or even a pen to use that system. The three-column Dean One-Page Plan will work with a sharpened stick and dirt. If you're rebuilding society in a post-thermonuclear war scenario, you might have to change the second column from "BUY" to "Barter," and the third column from "CALL" to "Visit." (Or perhaps change the third column to "Annihilate with rocks and sticks" if humanity repeats itself with its last chance to start over. And I think it probably would.)

Here's a cute video about Internet addiction, from the WhatTheBuckShow. It's tongue-in-cheek, but it does provide a bit of information about the problem at hand:

www.youtube.com/watch?v=nXpYUkT5gjo (URL 12.25)

Backing Up

Back up your work frequently. Nothing is less effective for time management than doing hours (or years) of work and losing it into the air. Computers crash. Files become corrupted. Hard drives fail. And in my 17 years of daily computer use, I will tell you from experience: It's not a matter of *if*; it's a matter of *when*.

I save documents as I go. I turn off autosave in the properties of any program I work frequently with and save manually every few minutes, using the Ctrl+S keyboard command. Autosave can be obtrusive. It just gets in the way of my flow, because I have to wait for the autosave to complete before I can type anything else. But you probably shouldn't turn off autosave unless you're going to be fairly obsessive about saving manually. Like me.

I'll save even more frequently than every few minutes if I'm on a roll. If I type a particularly brilliant sentence, I'll save. If I make a really cool edit in my video-editing program, I'll save. It becomes second nature to do this if you work at it. Train yourself to save as you go.

I back up a text document as soon as I'm done working on it for that particular session. I'll either back it up to a USB thumb drive (that I carry with me most everywhere, on my key ring) or email it to myself at a free Gmail account that I use only for backups. With larger files, I back them up to a removable hard drive at the end of a bit of work, and once a week I back up everything to a 500 GB FireWire drive. It seems like a lot of work, it seems a little obsessive-compulsive, but in my computing history, in the literally millions of files I've created and edited, I've lost only one, and it wasn't a very important one.

Be sure you have good antivirus software with up-to-date virus profiles. And run scans once a week. I generally make Thursday afternoon my day to scan and then back everything up on all three of my computers. It takes a bit of work, but from a time management standpoint, there's no substitute for this kind of care. If your life exists mostly on computers, it's good to keep your computers happy, humming, and backed up.

Thursday is also my day to blow my computers clean with canned air, vacuum out the fans in my server computer, and even vacuum the house, as well as print out my new Dean One-Page Plan page. Thursday feels like Sunday to me, the day before the beginning of my week. I don't know why, but that's the way my internal clock works; I like to get everything done on the same day so I can start fresh the next day.

Keep on Time-Managing On

You'll figure out your own day, and your own way, but I truly believe that some sort of time management system is needed to do the amazing amounts of steady work that are required to make a dent on YouTube, as well as in the world, these days.

13

Becoming a Success Story

Alan Lastufka

Building Your Brand

YouTube is a great jumping-off point. But if your goal is fame, you will eventually have to step beyond the 'Tube. Unless your videos routinely receive millions of views, you won't be earning enough to put your kids through rehab, and you most likely won't be recognized on the street (unless you happen to be at a YouTube gathering). YouTube is a great way to get your foot in the door, it's a great way to network, and if you want, it can be a great way to make a bit of *passive income*.

PASSIVE INCOME

Passive income (or, as Michael calls it, ***mailbox money***) is money that is earned even while you're not actively working. While you're at your day job, you're paid only for the actual hours you're working. When you're not there, you're not earning anything. Making a video may take anywhere from one to five hours, but once monetized, that video earns you money for weeks, maybe even months, as long as it's still receiving views. Repeat this process, video after video, and you begin earning money while you're sleeping, while you're out walking the dog, while you're at your day job, and so on. Passive income is a good thing.

Many of the people who are successful with YouTube have realized it's a vehicle for building their brand. So, what's their brand? It's their *name*.

Websites come and go. YouTube may not even exist a year from now or three years from now. So, building your home on YouTube assumes a certain amount of risk that at any time, for any reason, Google can take that home away from you. But if you build your own site, independent of YouTube, you are back in control. You're in control of content and which advertisements are shown on your site. You choose how you communicate with your viewers, whether it is through a forum or a streaming video chat.

Michael taught me this lesson early on, and now I'm passing it on to all of you. While subscribers are important, they are not the end goal. You want viewers who will know where to find you should YouTube.com return a 404 error message one day.

So, where to begin? You have numerous options for expanding your presence beyond YouTube, but here are a few of the more popular choices:

- **You could start with something as simple as a blog**: Many free blogging services exist, and we've even listed some of the better ones in the appendix. Blogs are easy to set up and maintain. Your blog may consist of text posts or embedded YouTube videos (just remember that these videos would also disappear should YouTube vanish). Most blogging services also allow you to run ads on your posts, meaning you'll be earning even more of that valuable passive income.

- **You could create your own social network**: That might sound scary, but it can be done, and you don't even need to know what HTML stands for to do it! Numerous social network–building sites and scripts are available, and most of them are free. Many YouTubers—including LisaNova, *www.youtube.com/lisanova* (URL 13.1), and the vlogbrothers, *www.youtube.com/vlogbrothers* (URL 13.2, both of whom are interviewed later in this book—use the popular social network–building site Ning at *www.ning.com* (URL 13.3). Ning allows you to build a site where your viewers can create their own profiles and upload their own videos, photos, and music; it's much like your own personal Facebook or MySpace, consisting only of your viewers and friends.

- **Start your own TV show**: Okay, maybe you've noticed that I have ordered this list easiest to hardest, but it's not impossible. Michael Buckley (*www.youtube.com/whatthebuckshow*; URL 13.4) used to broadcast his videos on public-access TV. If YouTube were to have failed back then, he not only would have had his extremely popular blog to interact with his viewers, but he would have had his TV show as an outlet too. He now occasionally shows up on FOX News offering a rapid-fire minute of celebrity updates. Hank Green of the eco-minded vlogbrothers has appeared numerous times on the Weather Channel. The point is, if YouTube disappeared, many of us would still know where to find our favorite YouTubers. If YouTube failed tomorrow, would your viewers know where to find you?

My co-author Michael had a cable-access show on five Southern California TV stations for a while, called *Stink Fight, Radio on TV*. Some of the show excerpts are on his YouTube channel, and he wrote an article for the O'Reilly Digital Media site explaining how to get a show on cable access.

www.oreillynet.com/pub/a/oreilly/digitalmedia/2008/03/06/put-your-photos-on-tv-pt2.html (URL 13.5)

> **Note** When you're posting videos on YouTube or creating your own online realm, you might want to use your actual name as your username on some or all sites so that your fans can always find you. Or register your username as a domain name, and use that for your external site.

Michael Buckley was also smart enough to register an account with another video-sharing website. YouTube may be the biggest and best, but YouTube is also a huge pond where new users begin as small, tiny, powerless fish. Registering and focusing on a smaller video-sharing website might be better suited for those who'd prefer to be a bigger fish in a smaller pond. Or, in Michael Buckley's case (on Revver.com), it was simply a great backup plan.

YOUTUBERS ON TV

You don't have to land a regular spot on an entertainment show or start your own show to be on TV; several YouTubers have been interviewed, spotlighted, or otherwise recognized on television for their success with online video. NBC used to run a series called *iCaught*, which I recorded a segment for, that reported on popular online video trends and interviewed the creators behind those trends. Political vloggers have been featured on CNN, while amateur directors have had their YouTube videos air on the Independent Film Channel (IFC). My YouTube video *Five Stories* (*www.youtube.com/watch?v=wsF4zOYE__g*; URL 13.6) aired four times in May 2007 on IFC. It's all about making quality videos and being easy to contact (in other words, include an email address in your channel description).

In the next section, you will learn about some popular alternatives to YouTube. All these sites can be used in conjunction with, or instead of, YouTube. Each has its own pros and cons, and some pay, while some don't. Some have awesome distribution networks, or large built-in audiences; others require you to drive your own traffic to your videos. But if you simply can't find traction on YouTube or you want your own backup plan, check out some of the following sites.

Revver.com

Revver (see Figure 13-1) was the first video-sharing website I registered with after YouTube. Revver pays *everyone* who uploads a video, unlike YouTube, which, as we know, requires you to apply to be (and be approved as) a Partner and which can, at any time, revoke the partnership privileges. Because Revver pays all video uploaders, every video is prescreened by Revver's team of editors before it goes live. This means every time you upload a video, it may be anywhere from a few minutes to a few hours before you see it live on the site or are able to share it. The prescreening team looks for any copyright violations, porn, or hate speech, none of which is allowed on Revver.

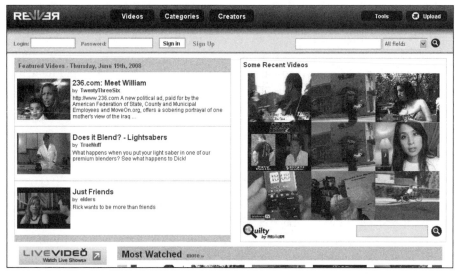

Figure 13-1. Revver's front page.

Revver pays less than YouTube from the stats I've been able to check (both my and LisaNova's views versus earnings there). However, Revver allows you (and the editors) to give your video a rating, much like the Motion Picture Association of America (MPAA) does with feature films. This means if you have a video with a little cursing or violence, you can tell Revver at the time you upload to rate the video 17+ or Explicit. These are the ratings currently available on Revver:

General, for all audiences

13+, for mild cursing or violence

17+, for if you drop the f-bomb

Explicit, for videos that should be viewed only by adults. (Again, Revver does *not* allow porn.)

If the Revver screeners agree with your rating, they will let it stand, but if they disagree, they are allowed to upgrade or downgrade the rating.

While Revver's "pay everyone" policy is attractive and the rating system is useful for those who worry about flagged and removed videos, Revver has a noticeably lower viewership than YouTube. If you don't have another vehicle for driving traffic to your videos, such as a popular blog, a few thousand MySpace friends, or a healthy Digg account, your video probably won't break the 100-views mark.

Revver currently does not offer any live video-streaming (watch-while-it-happens) services.

Livevideo.com

Livevideo (see Figure 13-2) has become the home for banned YouTube users. Live video was dubbed "The Promised Land" by many YouTubers after a number of high-profile YouTubers decided to move from YouTube to Livevideo. It's believed Livevideo paid many of these early vloggers to make the switch. However, without exception, every one of those vloggers eventually returned to YouTube.

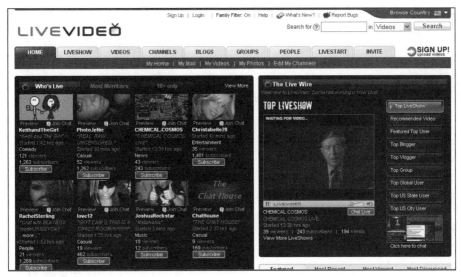

Figure 13-2. Livevideo's front page.

Livevideo, despite its name, wasn't actually a live video-streaming service until very recently. Its first few years in service were as a video-sharing site only, much like YouTube. Recently, Livevideo added and heavily promoted its live video-streaming features, attempting to compete with sites like BlogTV and Stickam. It's too early to say, as of this writing, whether its service offers any benefits over its competitors.

Livevideo does not publicly pay any of its users, though deals have been rumored and hinted at, as mentioned earlier. Many YouTubers keep semi-active or active Livevideo accounts, simply as an additional outlet for their videos or to offer a different variety of video than what they upload to YouTube.

Metacafe.com

Metacafe (see Figure 13-3) is another video-sharing site. Like Revver, Metacafe allows for all users to potentially get paid (by enrolling in the Producer Rewards program, which I'll explain in a minute) and does not offer any live video-streaming services. Metacafe's Producer Rewards program is its version of YouTube's Partnership Program and is available on a *per-video* basis, rather than a per-channel basis like at YouTube. You submit each of your videos into the Producer Rewards program (which requires you to agree to a few additional terms of use), and then each video is accepted or rejected on an individual basis.

Figure 13-3. Metacafe's front page.

Videos enrolled in the Producer Rewards program display a little icon indicating such, and they also display the amount of money that video has earned in an attempt to encourage others to use the program. Metacafe is the only website I'm aware of that publicly displays the dollar amount each video and user has accumulated on the site. From the public numbers available on Metacafe and my own experience with Revver, it appears Metacafe pays a bit better per view.

Meanwhile, Back at YouTube

You can choose from hundreds of other video-sharing websites; those listed here are just the few I've had personal experience with. It can't hurt to mirror your videos on these sites. Again, on these sites, you'll be a much bigger fish in a much smaller pond than you would be on YouTube, but with that comes a much smaller built-in audience, and you'll have to generate most of your views yourself.

YouTube is, by far, the number-one video-sharing site. According to Wikipedia:

> "In January 2008 alone, nearly 79 million users watched over 3 billion videos on YouTube."

YouTube is a cultural force and a phenomenon, not just a website. It's mentioned in popular hip-hop songs (Mariah Carey's "Touch My Body") and popular sitcoms (*South Park*, *The Simpsons*, *Family Guy*, and others). It draws sponsors and advertisers on par with network television, with most companies running special video-oriented contests, giving away money and prizes to YouTube users. And it creates memes that spin off from just a YouTube video to the real world.

Memes That Have Jumped the 'Tube

Chris Crocker. "Chocolate Rain." *Star Wars Kid*. These clips didn't just go viral; they *made* celebrities. Chris Crocker turned on his camera and screamed for us all to "leave Britney alone!" His clip was later used in the film *Meet the Spartans* from 20th Century Fox, and he appeared on numerous daytime talk shows.

MEMES

According to Wikipedia, an Internet *meme* is "a neologism used to describe a catchphrase or concept that spreads in a fast way from person to person via the Internet." Memes generally can't be planned or faked; either an audience is going to enjoy, share, and parody your video, or they're not.

Tay Zonday wrote and recorded "Chocolate Rain" (*www.youtube.com/ watch?v=EwTZ2xpQwpA*; URL 13.7) in his living room, just as he had a dozen other original songs before uploading them to YouTube. But unlike his other music videos, "Chocolate Rain" struck a chord, inspired numerous parodies (so many in fact that for one day YouTube featured not only the original "Chocolate Rain" video from Tay Zonday but also 11 parodies of it on the front page) and landed Zonday a commercial deal with Dr. Pepper.

Zonday was invited to play live on several late night shows, including Jimmy Kimmel, where he performed the song live. Zonday's Dr. Pepper ad was for the product launch of Cherry Chocolate Dr. Pepper, a new flavor created in honor of the song (!).

The *Star Wars Kid*, a clip of a teenage boy waving what looks like a broomstick around like a light saber in his garage, went on to be parodied in Weird Al's "White & Nerdy" music video (*www.youtube.com/watch?v=-xEzGIuY7kw*; URL 13.8), as well as garner countless other popular media references.

These are just a handful of examples. Most memes (like viral videos) happen by accident. They are embarrassing moments caught on tape (the grape-stomping lady and Miss Carolina and her maps) or private moments that were never meant to be shared (the Paris Hilton sex tape). Memes are rarely planned and are most effective when they're organic. If you're interested in the history of online video memes, Wikipedia has collected a page of them where they are listed chronologically.

http://en.wikipedia.org/wiki/List_of_Internet_phenomena (URL 13.9)

Weezer's "Pork and Beans"

Grunge-rock group Weezer appears to have wasted more than a few nights watching videos on YouTube; their music video for the single "Pork and Beans" (*www.youtube. com/watch?v=muP9eH2p2PI*; URL 13.10), featured famous and infamous YouTube memes and their creators, including Crocker and Zonday. The video takes you from the band interacting with one YouTube meme to the next. Numa Numa, Miss Carolina, Will It Blend, Dramatic Chipmunk, and so many others lip-sync along to the catchy track as the band plays in the background.

The "Pork and Beans" song was featured on the front page but long before that, Weezer front man Rivers Cuomo had interacted with the YouTube community via a personal vlogging channel, riverscuomoalone (*www.youtube.com/riverscuomoalone*; URL 13.11). On his personal vlogging channel, Cuomo invited viewers and fans to help him write and record a song from scratch, using the video comment section to conduct informal polls and incorporating the most popular ideas into the song in progress. It's the first time I've seen a band get that involved and interact with their fans without it being some marketing ploy.

Education and Charity via YouTube

Being successful isn't just about subscribers and view counts; Internet celebrities can give back or donate their time in numerous ways so they can, you know, still sleep at night. While making shout-out videos for your friends and early subscribers is nice and does give back to the YouTube community, remember that there is a whole world outside YouTube, filled with people who need more than a boost in channel views.

How-Tos and Documentaries

For those YouTubers who possess a certain skill, like being fluent in Photoshop, there is an entire category on YouTube dedicated to how-to videos. From simple tutorials and demonstrations to ongoing documentaries, the How-To and Education categories offer a wealth of knowledge in short, easy-to-digest videos.

IceflowStudios (*www.youtube.com/iceflowstudios*; URL 13.12) is one of my favorite examples of a YouTuber who mostly uses his channel for tutorials. IceflowStudios focuses mainly on Photoshop tutorials, with each episode demonstrating, step-by-step, a certain technique or tool. IceflowStudios' tutorials are always very thorough, have excellent video and sound quality, and offer interesting projects that you'd actually find useful while following along at home. And the host is only 21 years old. He records his professional-looking tutorials in his bedroom on a Mac laptop, proving you don't have to be a big production company or established charity to run a company that helps others.

A really amazing how-to series that went viral is *You Suck at Photoshop*.

> *www.youtube.com/watch?v=U_X5uR7VC4M* (URL 13.13)

This series offers really useful Photoshop lessons but also diagrams the creator's divorce, life falling apart, and generally hilarious passive-aggressive outlook on life. There has been much speculation as to whether the life stuff he mentions is staged or scripted, and there is a lot of mystery about who the guy actually is. Some think it's comedian Dane Cook. The voice certainly sounds a lot like Cook, and it's reminiscent of his style of humor.

The Uncultured Project

While a graduate student at the University of Notre Dame, Shawn Ahmed, who goes by the YouTube username UnculturedProject, *www.youtube.com/unculturedproject* (URL 13.14), put grad school on hold to fly to Bangladesh (where his parents emigrated from many years ago) to do some charity work (Figure 13-4). Using his laptop, camcorder, and mobile phone, he broadcast it all on YouTube. I contacted Shawn in Bangladesh and had the opportunity to ask a few questions.

Figure 13-4. Shawn Ahmed outside the home of a family he helped in the Tangail district.

Alan: What is the Uncultured Project and why on YouTube?

Shawn: The Uncultured Project is something I came up with after thinking about how to make important social issues, like global poverty, accessible to people in my generation—that is, the generation that tends to watch YouTube more than TV or movies. I love YouTube, but when it came to social issues like global poverty, there was nothing more than what you'd find on TV—commercials and documentaries. I guess this is because when it came to videos about global poverty, unlike comedy videos or videos about celebrity gossip, YouTube seemed to be dominated by big-name organizations.

Alan: So how does your approach differ?

Shawn: When it came to getting started, I approached this like a vlogger. But, unlike a regular vlog, I knew there would be a high cost of entry (buying a laptop, putting school on hold, moving to a different country, as compared to other vloggers whose cost of entry is simply a $39 webcam). I could talk about celebrities anywhere in the world from my home. But, if I just talked about poverty from my home, it would be nothing but depressing facts, figures, and stock footage.

YouTube is about making things personal, accessible, and the videos are usually positive. That's what I kept in mind when starting this project. And, as I mention on my channel's home page, to start this project, I had to withdraw from graduate school in good standing, pack my bags, grab my computer, buy a camcorder, and fly to Bangladesh. The rest is history...and history in the making.

Project for Awesome

On December 17, 2007, popular YouTubers John and Hank Green (vlogbrothers), *www.youtube.com/vlogbrothers* (URL 13.15) organized a charity project they called the Project for Awesome. The goal was to involve as many of their viewers as possible in making videos that spotlighted various charities and then comment spam each other's videos to the top of the lists. As we've previously discussed, videos get 48 hours before they fall off the YouTube lists, so to register the most effective first wave of commenting, Hank and John organized those who wanted to participate via a secret emailing list. Each video made for this project contained the same thumbnail, an image I designed for the project (see Figure 13-5). This image was distributed to everyone who wanted to participate prior to the project.

Figure 13-5. Project for Awesome's official video thumbnail, designed by Alan.

THE ECHO PARK FILM CENTER

I participated in the vlogbrothers' Project for Awesome, not only as the person who designed the thumbnail we all used in our videos, but I also made a video and spotlighted a charity. The Echo Park Film Center is located in Los Angeles and provides video equipment and multimedia education to children and young adults who would not otherwise have access to these creative tools and knowledge. For more information, visit the center's official site at *www.echoparkfilmcenter.org* (URL 13.16).

You can see our Echo Park Film Center video for Project for Awesome here:

www.youtube.com/watch?v=WCxpNg6Ksbk (URL 13.17)

Michael Dean hooked me up with them, helped me obtain images, did the music for the video, and offered feedback on the editing and audio from my rough cut that made it into the final cut. This is one of the many projects we've done together.

Hank and John set a specific day and time for us all to upload (so we'd rise and fall off the YouTube charts at approximately the same time), and as the first few videos went live, they used their secret emailing list to spread the YouTube video links. Hundreds of viewers took part that day (some proudly skipping school and work, myself included, to participate), leaving tens of thousands of comments on participating videos. Hank and John's plan was a success; at one point during the first 48 hours, our videos (and their identical thumbnails) landed 19 of the top 20 slots on the top 100 Most Discussed videos for the day.

Other YouTubers outside the project's secret emailing list quickly took notice and began making their own charity videos for the event. The project was covered in numerous online articles and eventually had its own song written about it, called "The Day That the Nerds Took Over YouTube."

www.youtube.com/watch?v=W3HqLDS-ri8 (URL 13.18)

More than 300 charities were promoted across hundreds of videos, and one can only imagine the dollar amount that was raised by all of those donating. Other YouTubers have supported various fundraising projects via YouTube, some even going live on Stickam or BlogTV for 24 hours straight to raise money and educate viewers about their charity. You can use your audience for good in many ways, once you've built it.

Note Editor's Note: We had a great sidebar by Michael on Professionalism. but Michael and Alan are so darned professional they over-wrote this book and that sidebar had to be cut for length. They've posted the whole inspirational piece on Alan's site and we highly recommend you read it to be a success story on the Tube (or off).
http://viralvideowannabe.com/professionalism (URL 13.19)

Hacking Embedded Videos to Autoplay

As you increase your web presence, you may want to embed your YouTube videos on external pages. Perhaps you'd like to include your latest video in your new blog post or even include it on your MySpace profile. That's great, and I encourage it. Embedding your video on an *external* page (a page outside YouTube) gives visitors to other websites the chance to watch your video without leaving that site. However, viewers of embedded videos won't be able to rate or comment on that video, and they won't be able to subscribe from the external site, something to consider before embedding.

Embedding is fairly straightforward. YouTube provides a string of code next to each video that you can copy and paste into your blog or MySpace profile. Once it's pasted there, your video appears like magic. By default, the video appears in a YouTube player with a large Play button in the middle of the screen. For some instances, though, you may want your video to play automatically without requiring your viewers to click that Play button.

Here's how to hack the embedding code to *autoplay* (automatically start) your video:

1. Start with the default embed code, provided by YouTube:

```
<object width="425" height="344"><param name="movie" value="www.youtube.com/
v/717oxEOfERQ&hl=en"></param><embed src="www.youtube.com/v/717oxEOfERQ&hl=en"
type="application/x-shockwave-flash" width="425" height="344"></embed></object>
```

2. Find the video URL; it will appear in the string of code twice and will look something like this:

```
www.youtube.com/v/717oxEOfERQ&hl=en
```

3. Add &autoplay=1 to the end of the video URL like this:

```
www.youtube.com/v/717oxEOfERQ&hl=en&autoplay=1
```

4. When you're done adding &autoplay=1 to the end of the video URL both times it appears in the embed code, you're finished. Your final embed code should look like:

```
<object width="425" height="344"><param name="movie" value="www.youtube.
com/v/717oxEOfERQ&hl=en&autoplay=1"></param><embed src="www.youtube.com/
v/717oxEOfERQ&hl=en&autoplay=1" type="application/x-shockwave-flash" width="425"
height="344"></embed></object>
```

Now your video will automatically start each time the web page it is embedded in loads. Autoplaying is a great way to save your visitors the hassle of clicking the Play button. Just be careful to not embed more than one autoplaying video on any single page. As a viewer, nothing is more irritating than trying to pause multiple players that have all started playing different videos at the same time. It can crash people's browsers, which will lose you the viewers you're trying to gain. Not many people have the attention span to try to view a new site over and over. They'll simply go someplace else.

We mentioned the oversaturation of things like autoplaying videos and music and MIDI songs and glitter text on your profiles earlier in the book, but it bears repeating. Be prudent. Be classy. Be minimalist.

The Moral of the Story

The moral of the story is build your brand. You want to make sure your audience knows you, not just one video of yours, because that video might disappear some day. You want to diversify your online presence; start a podcast, write an eBook, or host live events and stream them online. You'll have plenty of opportunities to interact with your friends and viewers and make, or even donate, a little money in the process.

14

Closing Arguments

Michael W. Dean

This chapter is a veritable manifesto on life. If you're wondering, "What the heck is this doing in a tech book?" well, feel free to ignore it. Alan and I have already delivered on the promise of the book title in the first 13 chapters, so consider this a free extra. But we believe that every how-to book should contain some "Why?" to make sure you use your new skills for good, never for evil.

Here you'll find words of encouragement, some sidewalk Zen on the art of living life (online and off), some talk about new things you can try going forward, and a little bit of a good ol' punk rock pep talk. This is the commencement speech that takes what you're doing on the 'Tube and rockets it beyond…into the future of your life, and the lives of every citizen of the planet Earth.

Alan and I talked a little before about "not making YouTube your world." I want to tell you why you shouldn't and, moreover, how you can change the world and become part of the global community, using YouTube as a jumping-off point.

What's Wrong with the Internet

You know how in every horror movie and sci-fi movie, one character is the *portent*? This is the grizzled old miner who says, "You youngsters shouldn't go camping out there tonight. It's the anniversary of the night them other kids went missing. Folks 'round these parts say a witch done ate 'em!" Or it's the crazy scientist who says, "We should listen to these signals from space! They're a warning! These aliens are trying to tell us to change our ways! And our DNA!" In this chapter, that old guy is me.

The portent seems like a nut in the first act, but by the third act, all the characters, and the audience, are saying "Maybe that crazy old guy was right!"

Gather 'round the fire, kick up your feet, put on your "open-mindedness hats," and listen while "Ol' Grandpa Punk" regales you with tales about "When I was your age…," what that means now, and how you can use it going forward. Since I'm old and you're not, you're going to be alive long after I'm worm food. And I don't have any kids, so I'd love to pass something on to somebody.

The Problems with Social Networking

Social networking provides websites to people who shouldn't be allowed anywhere near a computer.

Social networking is trumpeted as the "wave of the future" by companies like News Corporation (which owns many magazines, newspapers, and Fox Broadcasting Company). It's "the next big thing" because companies who own the old media now also own the new media. For instance, Fox owns MySpace.

Social networking is big with companies because the users create the content for free. They don't have to be paid. That's the boon to corporations in Web 2.0. Back in Web 1.0, companies paid people to create the content. I worked at a company in 1997 in San Francisco called catalogs2go.com. The company paid me (and about 50 other writers) 30 bucks an hour, full-time, to read printed catalogs and summarize them for its up-and-coming website, which never launched. The company went under before it went online. The domain catalogs2go.com is currently for sale, and it has been for sale for a decade. catalogs2go.com was not alone in this dot-bomb bust. Thousands of companies lost millions of dollars each. It was the end of Web 1.0.

In Web 2.0, you create the content. Companies have to pay only programmers, not content creators, and much of the programming work is outsourced to India at 1/10th the cost of work done in the United States. In Web 2.0, your brilliant work is competing with millions of other people producing brilliant work, all barking to be heard, while you have to juggle with the terms of service imposed by the company to keep all those users somewhat civilized.

I also don't like business-oriented social networking sites. I have no need for them. I'm a well-employed writer and professional filmmaker. I get more work offers than I can take on, and I do not have a profile on LinkedIn, IMDB Pro, or any similar sites. If people want to find me, they can. Start on Google, and within a few clicks, you've got my email address. This is how I get a lot of the freelance work I do…people searching me and finding me. You can learn to be this much in demand too.

Everything you do, in any medium, for free or for pay, helps increase the value of your "brand," that is, the value of your name and your resumé as someone people want to hire. This is why I spend most of my time online working on my own sites, not on other people's sites. I own my sites. They can't go away and I control the terms of service. My blogs are hosted on my own server, not on some social networking blogging site, as are my podcasts and most of my film and book information. I can't be banned. I can, however, ban people who irk me.

Why Kant Johnny Writ?

I've heard it said that the Internet is responsible for a decline in literacy—that people don't bother with proper spelling and punctuation because the Internet is so immediate. Statistics actually show that overall literacy rates are not going down. What I think is happening is that we're just seeing more poor writing because of the Internet.

Back when it cost money to *publish*, to print words on paper, all written media was authored by professionals, edited by professionals, and published and marketed by professionals. It narrowed the field of what was available, but at least the technical quality was decent. Ever since the World Wide Web debuted in 1992, most writing is written by anybody, edited by nobody, and published by a mouse click. Just because everyone *can* have their thoughts read around the world doesn't mean they *should*. Or at least I don't need to read them.

Web 2.0 has an amazing potential to unite, educate, and foster change, but it's been cut off at the knees by corporations who want to remove humans from the marketing equation and just have machines automatically dock credits from our bank accounts. They want to pick pockets by remote control. At the rate things are going, eventually the buyer (or "victim") will be eliminated.

What's Right with the Internet

YouTube is an exception to my "social networking is useless" rule, because it actually provides a forum for citizen media and gives a valid voice to people in a way that the masses can find easily and pay attention to. But if you're going to be on YouTube, I certainly recommend having your own website elsewhere and linking it on your user profile and on the More Info link of every video you make. Anything can go under, even something as solid and popular as YouTube. (As an old fart of the Internet, I've seen a lot of big things come and go, for a lot of different reasons.) You want people to know where to find you, and you want people to have gone to your site before so they can find you again if the site linking you goes down for good.

I love other things about the Internet. It enables me to work at home. I don't have to go to an office every day. I get up when I want to, work when I want to, and live a life I've wanted to live my whole life. I've consciously structured my decisions over the past decade to end up where I am now. However, living like this not only takes a lot of skill and an inordinate amount of work, but it also requires a lot of discipline. Most people, if they don't have a boss telling them when to work, simply don't. I work about 80 hours a week, I work on weekends, and I often work on my birthday and on Christmas.

A lot of making a living at art involves being on a mission—being absolutely sure in your heart of hearts that everything you do, from taking on a huge project that lasts years to answering your first-ever fan email and every other email you ever get, is for a reason. You *can* make a difference if you examine your reasons for what you do and how you go about doing what you do as a result.

The Media

The problem with big media is that there is no such thing as "news" anymore. Most stories reported as "news" are about things that are owned by the corporations doing the reporting. This covers the gamut of everything from social networking sites to theme parks, TV shows, movies, rock bands, and even politicians. This also includes the things that the media chooses not to report. They tend to not report things that they are in competition with, or things that are critical of things they own…including politicians.

What's right with the media is that a tiny percentage of it is actually great. I recommend watching the small part of TV that is great. Here's some of what I like:

C-SPAN

C-SPAN shows what's *really* going on in the world, "behind the curtain" of politics. Watch C-SPAN. It's the only truly honest media out there, and there are no commercials! If you're concerned that America is being sold out before your eyes and you wish there were a few heroes trying to stop this from happening, you can watch all that and more, *live*, from the floors of the Senate and the House of Representatives.

- The official YouTube channel is located at *www.youtube.com/user/CSPAN* (URL 14.1).

BREAKING NEWS

About 24 hours after I typed the preceding paragraph, I was watching C-SPAN and saw Rep. Dennis Kucinich on the Congressional floor presenting his unannounced resolution calling for the impeachment of President Bush. Kucinich spent nearly 5 hours reading a list of 35 charges, mostly Constitutional violations and Geneva Convention violations. If you missed it, you can see some of it here:

www.youtube.com/watch?v=1qy3z7XWtQc (URL 14.2)

Regardless of how you feel about it, it was history in the making, and most Americans missed it because most Americans don't watch C-SPAN. I was also a little shocked to note during the following couple of days that the mainstream media had pretty much ignored the event. Regardless of how you feel about it, it is news—but apparently not to major news outlets. They were busy talking about Obama's flag pin, McCain's haircut, and who won this week on *American Idol*.

I, however, did blog about the impeachment proceedings, while it was actually happening:

www.stinkfight.com/2008/06/09/is-the-president-being-impeached/ (URL 14.3)

Even if you disagree with Kucinich's charges against Bush entirely, his right to stand up and read them *is* what the Founding Fathers were fighting to permit and protect. It was so beautiful to watch that it literally brought a tear to my eye.

If you don't receive C-SPAN in your cable package, a lot of C-SPAN media is already up on YouTube. Do a quick search. C-SPAN is one of the only, if not the only, mainstream media outlets that entirely allows noncommercial uploading of its content. You can find info on that here:

www.c-span.org/about/press/release.asp?code=video (URL 14.4)

But wait! There's more good stuff on TV wedged between all the crap:

- History Channel: Keep in mind with anything about history that "history is written by the victors." But hearing about history from the victors is a good place to start and better than not studying history at all. It's a great jumping-off point to see where to point your own extended research.

- Local and national news: Take it all with a pound of salt. And mute the commercials. They'll make your tummy hurt.

- *The Daily Show*: This show is amazing. (No official YouTube channel exists.)

- *The Colbert Report*: Colbert is wonderful. He's a complete parody of the "We'll feed you the news we want you to hear, and nothing true or important" mentality. And sometimes laughter is the best medicine. Or at least it's the best defense as the whole system teeters on the precipice. (No official YouTube channel exists.)

Most of it is up on YouTube too. (Shhhh! It's a secret!) And some of it isn't a secret. Many media outlets have their own official YouTube channels:

- National Geographic: Locate the official YouTube channel by going to: *www.youtube.com/user/NationalGeographic* (URL 14.5).

- The Documentary Channel is at: *http://www.youtube.com/user/DocChannel* (URL 14.6).

- The Discovery Networks: *http://www.youtube.com/user/discoverynetworks* (URL 14.7).

Balance your good TV watching with occasionally surfing all the rest of the channels, especially the ads, to see how horrible they are, to see the lies they tell, and to see how almost all of it is either partying or dying, interspersed with them trying to sell you crap that you don't need and make you think that if you don't buy it, you'll die sad and lonely.

Don't watch the media, be the media. And make it great. YouTube can be the media you utilize, or it can lead to jobs in conventional media. If you land one of these jobs, try to make a difference. They probably won't let you, unless you're sneaky and make light of the problems in the world (and in the media) in a way that makes the people who sign the checks think you're talking about someone else. But the joke's on them. You're really talking about all of it.

What's Wrong with the World

The 80s "Me generation" morphed into a post–Me generation. This is best summed up in a term I hear some variation of (at least in stance, if not in word) in almost every movie and in many music videos: "I'll bust a cap in your ass if you get in my way."

This "shoot to kill" attitude is influencing the rest of the world too, mainly through the media.

"I'll bust a cap in your ass if you get in my way" is also sometimes summed up as the softer but still aggressive "Don't you know who I *am*?" This phrase is usually shouted by at least one of the participants in any conflict, from a school-yard fistfight to a street gunfight to a world war.

Many of the most popular video games in the world, especially "Grand Theft Auto" and its clones, involve players pretending to shoot and kill people. I think you should question why the world is at a place where the popular games have you pretending to kill people.

The "I got to get mine," post–Me generation ideal doesn't exist just at the street level; the philosophy often guides corporations too. A corporation is simply a way of counting and distributing money and can be administered in a very ethical fashion, and this does happen. But corporations exist to make money, and if they are guided by people without ethics, they will make money at any human expense. We're seeing this in the collapse of the mortgage industry, and the Wall Street bailout. When my parents first married, a man working on the line in a factory could buy a house and feed a family. Now, families where both parents work can't make ends meet. Homelessness used to be mostly for drunks and the mentally ill. Today a lot of hardworking people are ending up homeless.

I used to think that "the Man" (the establishment) was intrinsically evil. That's an easy way to see it when you're young and don't really have much direct experience with government or big business. But in the past 15 years I've had a lot more direct dealings with the Man (which is not always men; it's a new era, and the Man is often comprised more or less evenly of men and women).

Now that I've worn a tie and worked in offices, paid my taxes, stood in line at the DMV, stood in line to get permission from the state to marry the woman of my dreams, and put a toe into the mainstream media distribution networks and cashed checks from them, here's what I've learned. *The Man is not so much evil as he is just boring.* Boring, square, and slow-moving, the Man really has no idea what people want. Not only is the Man boring and square, he thinks he's exciting and hip. Dealing with the Man will not so much crush your soul as make you cry tears of boredom and make you rush back to your bedroom to make some art that's interesting and share it with as many people as possible.

Pay attention, get involved, and learn what the issues are. They're incredibly complex. That's the idea behind our "representational democracy"—that most people are not smart enough to vote on the issues, so we vote in people who vote on the issues for us.

It may actually be true that most people are not smart enough, or informed enough, to directly participate in politics. But you don't have to be one of them.

What's Right with the World

People are not victims of circumstance nearly as much as they think. Rise above your problems and be great. Don't waste your six or seven decades doing nothing. Make a difference.

I always wondered when I was younger, "If the me now could meet me when I'm older, would I like him? Would he like me?" Well, yeah. We'd like each other. I do wish, however, that I'd learned from other people's mistakes, not just my own mistakes.

Some people lie so much that when they tell a lie, they think they're telling the truth. Learn how to filter through all the duplicity and remove those people from your list. Then sift through the rest to find the honest few who are both brilliant *and* easy to work with. Surround yourself with these rare humans, collaborate with them, befriend them, love them, and change the world with them, forever.

What You Can Do About All This

First, throw away the "bust a cap" attitude, if you have any. Everyone who has this attitude is, deep down, a scared little kid whistling in the dark. They're terrified of the world, are afraid to meet the world on realistic terms, and want to destroy what they can't control. Be confident, not cocky. Cocky will get you killed, jailed, or, at the very least, have you worried every day of your life, and worry makes you old quick.

Believe in the healing power of art karma. Have some social responsibility in the media you produce. Help others. Question authority, but don't blindly hate it. Learn about the politics and laws of your country and of other countries. If you live in the United States, memorize the Constitution of the United States (All law in America is based on it.)

If you're in a hurry, just memorize the Bill of Rights (the first 10 amendments to the Constitution). Did you know that most of the rights that are commonly violated are the rights that are enumerated in these first 10 amendments? And, most important, did you know that politicians currently in high office in this land treat the Constitution as something to be changed, and many people don't know the Constitution well enough to know when it's being changed? And did you know that most don't care and would rather watch *American Idol* or movies with lots of car crashes?

Note One of the many fun "gift to the world" little projects I've done with Debra Jean Dean was engineer and record high-quality files of her reading the Constitution, the Bill of Rights, and the Declaration of Independence. We released the files free to copy, under Creative Commons, so you can share them with anyone and use them in any way you please. Put them on your pod and listen to and repeat them until you know them. Then remix them in videos if you'd like.

Wired Magazine's "Geek Dad" column called them "A portable civics lesson." *http://blog.wired.com/geekdad/2008/01/ipod-plus-delca.html* (URL 14.8)

You can grab them here: *www.debrajeandean.com/* (URL 14.9)

Surviving and Thriving in the New Economy

High school doesn't teach you the things you need to know to survive in today's world. Futurist Ray Kurzweil pointed out in his book *The Age of Spiritual Machines: When Computers Exceed Human Intelligence* that school teaches you to show up for years at the same place at the same time each day and sit in little rows and do what you're told. (I highly recommend that book, *www.amazon.com/Age-Spiritual-Machines-Computers-Intelligence/dp/0140282025/* (URL 14.10).

Winning, or even surviving, in the economy of the present and the future does not involve showing up at the same place at the same time for 50 years. It involves small ad hoc companies that don't have to be in the same room (or even the same country) working intensely for a week, six months, a year, or two years, over the Internet, creating something great, quickly selling it (and outsourcing the manufacturing and distribution if it's a physical product), and then moving on to work intensely with other people on other projects. They don't teach that in school. In the old economy, an "expert" did one thing and did it very well. In the new economy, an "expert" needs to do many things well and be able to quickly and efficiently figure out new things and identify and solve new problems in new processes quickly and efficiently as they arise. The well-employed worker of the future is not a mighty giant. The well-employed worker of the future is a small, efficient, adaptable ninja.

Good further reading on this subject is *The World Is Flat 3.0: A Brief History of the Twenty-first Century* by Thomas Friedman: *www.amazon.com/World-Flat-3-0-History-Twenty-first/dp/0312425074/* (URL 14.11)

To make it in the coming times, you need people skills. You have to be able to work well with anyone, at least over the Internet and the telephone. And it helps if you speak more than one language, although you need to have a functional command of English, because that is the international language of commerce, science, law, and computing. (English is not the most common language on Earth. It's the fourth most common. However, English is by far the most common second language on Earth.)

You need to be able to sort (tell the truth from the lies, the useful from the useless), and utilize (decide in unique ways what to do with information, media, ideas, processes, software, movies, images, concepts, and even people). And you need to be able to do this while also quickly sorting though mountains of commercial and personal spam. Being active on YouTube is great training for some of these skills. But YouTube as training, in and of itself, is not enough. You need to read constantly, talk, learn, live, and love. And lose. And gain again.

It should go without saying, but you also need to be able to do basic math and read and write. And you need to use any old and new tools available to you, in basic ways, but also in new and unique ways.

Michael's Pearls of Wisdom:

- Never stop learning.

- Hone your skills and remain teachable.

- Prepare yourself with a wide set of skills.

- Be compassionate, and show that in your art.

- Conserve water. Use rechargeable batteries. Recycle.

- Ignore advertising.

- Become really good at whatever it is that you do that makes you unique.

- Be patient.

Jobs I had that I hated (digging ditches and telemarketing) not only made me a better person and gave me a better understanding of the world, but they also made me a better artist. Every skill and every experience arms you to better deal with every other task and every new experience. Don't live an isolated life, and try to learn from anything you see or do. And try to learn from everyone you meet, even if all you learn from that person is what *not* to do.

Making Lemonade from Lemons

Strife can create growth, so don't feel helpless when you experience bad times. I hate that fluffy greeting card sentiment "It's always darkest before the dawn," but there is some truth to it. Sometimes the greatest spiritual awakenings, even the kind that make sense to an agnostic (sometimes called "spiritual awakenings of the educational variety") come only after a great loss. After hitting bottom.

My daughter, Amelia Laine Worth (see Figure 14-1), died November 7, 2006, from leukemia. She was 22. She was a good artist and a great person, and she lost the luxury of time.

Figure 14-1. Michael W. Dean and Amelia Laine Worth. This was the day she got her black belt in karate, a few short years before she got sick.

Her death devastated me, and I thought I'd never climb out of that depression. But I did. A day doesn't go by that I don't miss her, but surviving the trauma of her death brought me some understanding of the world that I never had before. What's that saying? "You can't play the blues on a guitar that's never been in a pawn shop." It's true. I recorded a song for her and posted a video of myself singing it on YouTube.

www.youtube.com/watch?v=RqZNMO3oIdc (URL 14.12)

Doing this, and having her friends see it, was a small but important part of me healing some from the experience. I brought three positive lessons out of her death, after months of sleepless nights and more than a year of walking around in a daze, looking and feeling like a zombie:

- There is nothing that can happen to me, absolutely nothing, that I cannot get through. So, life is all gravy from here on out.

- There is absolutely no one alive anymore on the planet Earth who I am afraid of embarrassing, so I am going to do whatever I feel like doing. To live this way, you need to have a great internal sense of doing the right thing, and that comes with age, experience, and reflection. But I used to moderate my artistic output a bit, out of respect for my daughter. I probably didn't need to; she was a tough kid, very smart, and very accepting of opinions different from hers, but I did hold back a bit, for her. Now that she's gone, I am living in dedication to her by being even more of an unstoppable life force than before.

- A long life is not a guarantee. Amelia was healthy a couple years before she died, and she had several remissions when we thought she was in the clear. She even looked relatively healthy a few months before she passed (see Figure 14-2).

"Live each moment as if it's your last," because it might just be. And create every piece of art with meaning, as if it's the last thing you'll be remembered for. Because it might just be.

Figure 14-2. Michael W. Dean and Amelia Laine Worth, four months before she passed away.

Take your life experiences and reflect on them. Then reflect them back to the world, with your own unique additions, interpretations, and insight. Don't just make another cute kitty video or a nifty skateboarding accident reel. Have something to say, say it, and say it well.

You have the potential to change the world from your bedroom. Don't blow it.

Note The first draft of this chapter was a lot longer, but I trimmed it for length. To read what I cut, go to: *http://tinyurl.com/6n9kkf* (URL 14.13)

15

Interviews with Other YouTube Rock Stars

Alan is an expert on all things 'Tube. But different experts have different ways of doing things. So, we had Alan go out on a virtual field trip, imaginary microphone in hand, and hunt down some of the really big YouTube rock stars, corner them, and ask them how they do things.

The following people are interviewed in this chapter:

- Lisa Donovan (LisaNova)

- Hank Green (vlogbrothers)

- Michael Buckley (WhatTheBuckShow)

- Kevin Nalty (nalts)

- Liam Kyle Sullivan (liamkylesullivan)

 You can find more of Alan's interviews with other interesting YouTubers by going here:

http://viralvideowannabe.com/interviews (URL 15.1).

Interview with Lisa Donovan (LisaNova)

Lisa Donovan.

YouTube: *www.youtube.com/user/LisaNova* (URL 15.2)

Website: *www.lisanovalive.com* (URL 15.3)

LisaNova is the 24th Most Subscribed channel of all time on YouTube (as of this writing). She became a MADtv cast member as a result of her videos on YouTube. She has been featured in the *New York Times*, *AdWeek*, and *Wired* magazine.

Alan Lastufka: When and why did you start making videos for YouTube?

Lisa Donovan: I first started making videos on my LisaNova channel in June 2006. I had already been in Los Angeles for a couple of years and had ended up working in production on the MTV show *The Osbournes* along with other random production jobs. I eventually teamed up with director/producer Danny Zappin, and we formed a small production company called Zappin Productions at the end of 2005.

Our long-term goals of producing and directing independent feature films were temporarily put on hold as we found our time more and more consumed with doing corporate videos and working on other people's creative projects. Danny discovered YouTube a few months before I created the LisaNova channel. He encouraged me to start posting my own videos during our spare time between production projects. I really didn't know what kind of videos I wanted to make at the time and actually had a problem with just talking to the camera like the vloggers were doing. So, I just decided to make my first introduction video, *Introducing LisaNova*, a silent film with subtitles, French music, and an old film look.

Alan: How did you come to have your YouTube celebrity?

Lisa: My introduction video did pretty well. A fan base started to grow, and I was eventually featured on the home page twice in my first few months on the site. My P. Diddy video was featured on the YouTube home page the day that Google bought YouTube, and I got a lot of extra exposure from that. My *Teenie Weenie* video was also featured, and the casting director from MADtv saw it and called me in to audition for the show. I eventually was cast for four episodes at the end of that season and received a lot of press because I was the first YouTube person to land a real TV role. Besides the press, I've had quite a few videos that people really seemed to like, and my subscription base just kept consistently growing. I'm not universally known for any one hugely successful "viral" video. Instead, I'm probably known by many different people from a variety of different videos I've done. *LisaNova does YouTube* was a big boost mainly because of all the ridiculous drama it caused. Hundreds, if not thousands, of videos were made regarding the controversy surrounding all the comments we left on people's profiles leading up to the release of the video. It was quite an interesting social experiment, and it was a lot of fun to see all the interaction that took place because of it.

Alan: If you had one sentence to describe your YouTube channel, what would it be?

Lisa: I would say the LisaNova channel is a random collection of satirical videos about a wide variety of pop culture topics including celebrity spoofs, YouTube spoofs, and political satire as well as music videos and other stuff.

Alan: How has your success on YouTube helped your career outside of YouTube?

Lisa: I guess it has completely changed my career path. My YouTube success has definitely changed the focus of our production company and has led to many mainstream-media acting jobs and opportunities in TV and film. I still don't feel that Hollywood completely respects YouTube, but I think that will change in time. It's extremely exciting that people now have a way to build a fan base, make a living, or at least make a supplementary income by putting their videos on YouTube. It's cool to be able to build a fan base without having to deal with the normal Hollywood barriers.

Alan: Who is your target audience? And who is your actual audience?

Lisa: I think our target audience is college age and older, although I know a big percentage of our demographic is still younger than 18. The LisaNova channel probably has a much older demo than most of the top-subscribed channels, which appeal mainly to a younger demo.

Alan: In what ways do you interact with your audience?

Lisa: All the normal ways like comments and personal messages, and so on, as well as doing live vlogs/chats. I like to watch video responses when I have time and leave comments. I've met a few people in real life from that. It's always fun to meet people in real life who you know only as a username on YouTube.

Alan: Have your YouTube friends become your real life friends?

Lisa: A little bit, although most people I know from YouTube live in all different parts of the world, so that makes it a bit difficult to cultivate those relationships in real life.

Alan: Has interacting with people on YouTube made you more confident while interacting with people face-to-face on a daily basis?

Lisa: YouTube does help you build a thick skin. You have to be ready for massive rejection/criticism every time you put out a video. So, now I feel very immune to anyone's opinion of me. But I am a people person, and I have mostly felt confident in that regard. I love to meet new people and learn about their lives and their culture.

Alan: How much of your day is spent on YouTube?

Lisa: That really varies. There were long stretches where I would be on there 10 to 12 hours a day. Then there are times when I get burned out or really busy with other things and I barely go on at all. Typically, I'll spend at least an hour or so on YouTube on a normal day.

Alan: What advice do you have for up-and-coming YouTubers?

Lisa: It's very important to interact with the community, especially when you're first starting on the site. Subscribe and comment on other people's videos on the site, and make videos that occasionally have to do with other YouTube users or other YouTube videos (that is, video responses).

Quality and consistency are the other keys for building an audience. I would advise someone who's starting a YouTube channel to make several videos they feel good about before they start posting them. Make sure your videos have good sound and proper exposure and that they're something you like. Having a stockpile of solid videos can help you gain some momentum right out of the gate without taking a long break to shoot your next video. Comedy videos seem to play much better than others, and it's important to keep the videos short if possible because people have extremely small attention spans these days.

Alan: Could you repeat your success on YouTube if you started over today?

Lisa: It would probably be much harder for me to repeat the same level of success today if I were just now starting on YouTube. There's far more competition, and it's much harder to get seen these days. However, it's possible to blow up much faster now because there are so many more people here. I just think it's much harder to get to that point where you get mainstream exposure on the lists. For instance, it used to take 50 comments to be on the top 20 Most Discussed page. Now it takes about 1,000 comments, so it's much harder for a newcomer to get the exposure they used to get.

Interview with Hank Green (vlogbrothers)

Hank Green.

YouTube: *http://youtube.com/vlogbrothers* (URL 15.4)

Website: *www.ecogeek.org* (URL 15.5)

Hank and John Green (vlogbrothers), brothers for over 27 years, decided not to write, email, instant message, or text message each other for all of 2007 and instead make daily video blogs. Though the Brotherhood 2.0 project has ended, they update the YouTube channel at least once a week. The community of nerdfighters they helped create is now stronger than ever; it lives at *www.nerdfighters.com* (URL 15.6). The vlogbrothers' YouTube channel has been featured on NPR and in the *Wall Street Journal*.

Alan Lastufka: When and why did you start making videos for YouTube?

Hank Green: New Year's Day 2007. It was my first video I'd ever edited. And I had just signed on to do a full year of them...every other day. Man, that was dumb.

Alan: How did you come to have your YouTube celebrity?

Hank: I wrote a song about the last *Harry Potter* book the day before it came out. It got featured on the front page of YouTube. A tiny fraction of the million people who have watched that video became loyal viewers. And they told their friends, and we felt pressure to make better videos, and then more people responded to the better videos...and on and on.

Alan: If you had one sentence to describe your YouTube channel, what would it be?

Hank: We're just two dorky brothers who wanted to have a better relationship with each other and ended up having a pretty powerful relationship with 30,000 nerds.

Alan: How has your success on YouTube helped your career outside of YouTube?

Hank: Oh yeah, people are always being like, "Oh, that Hank guy is pretty cool, yeah." And then they find out about Brotherhood 2.0, and they're like "Hire him!" or "Bring him here to speak at our event" or "OMFG, I want to have his BABY!" That last one isn't really good for the career, but it's good for the ego. The biggest deal is that I got a gig making videos for Discovery Channel's new Green Cable Network. And, just in general, it's good personal branding. People just assume you're capable of anything, which is, of course, very far from the truth. But I don't generally tell them that.

Alan: Who is your target audience? And who is your actual audience?

Hank: Our target audience is anyone who believes that being yourself is cool and being smart isn't something to be ashamed of. Our actual audience is pretty much that, except it's mostly girls. About 75 percent of the people who watch us are female, which is really surprising. The only thing that would make us more interesting to girls than boys is our powerfully handsome faces, which is ridiculous.

Alan: In what ways do you interact with your audience?

Hank: There's probably a pretty sloped hierarchy here. There are the people who just watch who we never interact with. That is probably the biggest slice. And then there are people who comment, and we sometimes respond. But beyond that there's a select group of several thousand who we do tons of stuff with. We make music videos, we have scavenger hunts, they cover my songs, I cover their songs, we make collaborative videos, and sometimes we just sit around and talk. I hired a nerdfighter to work for my blog. We also do video chats sometimes, and we're going on a real, in-person tour, so we'll be hanging out with our audience face-to-face soon.

Alan: You refer to your viewers as *nerdfighters*. What are nerdfighters?

Hank: Instead of being made of bones and organs and stuff, a nerdfighter is made of *awesome*. Less technically, a nerdfighter is someone who believes in, and has no problems with, their dorky obsessions and interests and will never criticize anyone else for their obsessions and interests.

Alan: Have your YouTube friends become your real-life friends?

Hank: I'm not sure how to make the distinction between real-life and YouTube friends anymore. So, I guess the answer is yes.

Alan: Has interacting with people on YouTube made you more confident while interacting with people face-to-face on a daily basis?

Hank: Almost always. Sometimes I still get really nervous, but I think it happens less often now. Most people don't know what they sound like and even cringe a little when they hear themselves recorded. I know *exactly* what I sound like; I hear myself all the time. And I've even kinda started to like it.

Alan: How much of your day is spent on YouTube?

Hank: I really have no idea. Between one and six hours, I'd say.

Alan: What advice do you have for up-and-coming YouTubers?

Hank: Make videos you enjoy watching. I literally watch and enjoy watching my old videos. I sit there and laugh and say, "That was a funny joke." If you don't enjoy what you're doing—if you don't think it's cool or funny or worth watching after you're done—then no one else will. Also, it's a great cure for the "no one is watching, so why bother?" problem. If you're entertaining yourself, then at least you have one reason to keep doing it.

Alan: Do you feel you could repeat your success on YouTube if you started over today? Why or why not?

Hank: Maybe. There's no way to tell. YouTube changes constantly, so video makers have to change constantly as well. We have to keep on top of trends and make connections with people who are doing cool things. Plus, we'd need to have more good ideas, and you can never be sure a good idea is going to come by. I think it would be more difficult now. But there are still tons of people doing it, so it's certainly possible.

Interview with Michael Buckley (WhatTheBuckShow)

Michael Buckley

YouTube: *www.youtube.com/user/WHATTHEBUCKSHOW* (URL 15.7)

Website: *www.buckhollywood.com* (URL 15.8)

WhatTheBuck is the most popular entertainment show on YouTube, with more than 240,000 subscribers and 70 million views. Several original episodes are posted each week covering important topics like *Britney's Beaver* and *Tyra's Latest Weave*. He occasionally covers sports and politics but would much rather talk about what happened on *The Hills*.

Buck can also be seen often on the Fox News shows *RedEye* and *Lips & Ears*, can be seen live on BlogTV, and can be heard on several radio shows including Leeza Gibbons' syndicated radio show. In real life, he has a day job, enjoys playing with his four dogs, loves going to Broadway shows, and likes whitening his teeth.

As a result of his popularity on YouTube, he has signed a development deal with HBO.

Alan Lastufka: When and why did you start making videos for YouTube?

Michael Buckley: I didn't even know what YouTube was when my first video was posted. I was doing a show at my local public access station, and my cousin took one of the clips and posted it on YouTube. This continued for several months before

I started posting on my own channel. For the first few months, I really didn't grasp what YouTube was or all that would come of it for me.

Alan: How did you come to have your YouTube celebrity?

Michael: I slept with LisaNova! [laughs]

It was a lot of very hard work cranking out new videos on a regular basis. Throw in some good luck and networking with other users along the way, and I got to the level I now enjoy on the site.

Alan: If you had one sentence to describe your channel on YouTube, what would you say?

Michael: Um, one sentence? Me? Impossible. The WhatTheBuckShow channel is a four-eyed gay dude with a green-screen background making fun of celebrities. My channel is the anti–*Entertainment Tonight*; it's a show designed to take the piss out of celebrities.

Alan: How has your success on YouTube helped your career outside of YouTube?

Michael: My success on YouTube has launched a career for me that I otherwise would not have had. I didn't want to throw myself into the deep-yet-shallow pool of television commentators who were told what to say and how to say it. I wanted to create my own show with my own voice talking about the silly things that I wanted to talk about. I am lucky and grateful that doing this has unexpectedly become revenue generating and life changing. In terms of outside opportunities, I am a frequent guest on TV/radio shows, but I believe the Internet is where I will have my future and greatest successes.

Alan: Who is your target audience? And who is your actual audience?

Michael: To be honest, I try to appeal to everyone. I'm a people pleaser. I take pride in the fact that many people write to me and claim they never liked gay people or cared for celebrity gossip but they rather enjoy my show. My goal is to present material with humor, warmth, and joy so that even people who would not typically enjoy it find something about it to enjoy.

My actual audience is largely kids 13 to 18, which is funny because the show was not originally written for kids. But then I also have a lot of viewers who are 35 to 45. Being aware that young people are watching, I try to do at least one show a week about Hannah Montana, the Jonas Brothers, or something youth-oriented. They are the viewers who rate/comment/favorite your videos the most.

Alan: In what ways do you interact with your audience?

Michael: I make great efforts to reply to comments on my current video. I am very active on my website in the discussion forums, and I also do weekly live shows.

Alan: Has your circle of friends from YouTube transferred to your circle of friends in real life?

Michael: Yes, I am very close with people who I know from YouTube. Some I have met in real life; others I have not. Some of my closest friends I met on YouTube, and I view many as colleagues and friends. It is a strange new world that we are all figuring out as we go along!

Alan: Has interacting with people on YouTube made you more confident while interacting with people face-to-face on a daily basis?

Michael: I don't think so. I was always confident interacting with people. In some ways it has made me a bit shyer I think. Because my persona is so "out there" and loud on YouTube, I tend to be a bit more reserved in real life now.

Alan: How much of your day is spent on YouTube?

Michael: I am on all day. I may go a few minutes here and there without checking it, but if I'm off for more than an hour, it's only because I'm sleeping.

Alan: What advice do you have for up-and-coming YouTubers?

Michael: It is important to make videos you enjoy making. If you make videos trying to copy other people or have illusions of online fame, you will get discouraged and stop. Just like anything in life, make videos that you love, and others will see that and appreciate your efforts. In terms of practical advice, be visible on the site. Make sure you are rating/commenting/favoriting other people's videos to make it a more interactive experience. By doing this, you will build a larger audience for your own videos. Be nice and respectful of other users; you may think you're being funny, but mean comments will not get you respect or more viewers. Don't post one video and be upset no one sees it. You could post the best video in the history of YouTube and it gets 10 views; and that's just you checking to see whether anyone is watching. A lot of factors come into play when getting lots of views/subscribers on the site, so be patient! Keep making different videos, and one will catch on, or your diverse body of work will attract some attention. I also advise people to pull and repost videos that they are very proud of. If you posted a great video a year ago and it has a low view count, the chances are it's "dead" and will not catch on. If you are at a place where you now have maybe 1,000 or 10,000 subscribers and you posted a video you really love, repost it for your new larger audience to see.

Alan: Do you feel you could repeat your success on YouTube if you started over today? Why or why not?

Michael: I do believe I could repeat my success if I started over today. I always believed in myself and the videos I was putting out there. I worked hard to create and promote them, so I know I could do it again. I work even harder now to keep the quality up; knowing each day brings a whole new audience, I have to make a good first impression while maintaining the appeal to longtime viewers.

Alan: You host a live Q&A show an average of three times a week. Why do you feel this is important?

Michael: I think the most important thing I have done on YouTube is stay present. Many people just post videos and do not interact with the community. Many of them fade away. Some are lucky enough that their audience doesn't care. The moment I started vlogging and seeming more like a real person and not just some bitchy gay guy who made fun of celebrities is when I became more popular on the site. People like seeing the real person behind the videos. Doing the live shows/chats was just the next step in me being available and present to the people who enjoy my videos. I love doing the live shows; it's a rush and a pleasure to have the instant feedback and connection.

Alan: You write, edit, and perform all your videos alone, with numerous episodes produced each week without fail. How do you do all this while working full-time and balancing a home life?

Michael: It can be a challenge to make it all work. I've been doing it all for two years now, so I'm used to the heavy workload. And while writing/filming/editing the videos is a lot of work, it's also my passion, so I look forward to every one. If I feel overwhelmed, I take a week off or do fewer videos for a week, but I find when I am off, I miss it. It really nurtures my creative spirit.

My day job has been very supportive, and I have great benefits, weeks of paid vacation, and a 401(k), so it has been hard to leave! My husband has been very supportive and actually was the one who told me to have an opening credit with "Please Subscribe" and "Rate It If You Hate it," which has been very successful for me. And other 'Tubers use similar credits now too! He also gives me some jokes, but don't tell him I admitted that!

Interview with Kevin Nalty (Nalts)

YouTube: *www.youtube.com/user/nalts* (URL 15.9)

Blog: *www.willvideoforfood.com* (URL 15.10)

Kevin Nalty.

Kevin Nalty is one of the most-viewed YouTube comedians with more than 650 short online videos and more than 25 million views. He has been featured in the *Wall Street Journal*, *AdWeek*, and the *Los Angeles Times*, as well as on ABC News. He was honored at the 11th Annual Webby Awards. By day he's a marketing director at a Fortune 100 company. He speaks, writes, and consults about online marketing and viral video.

Alan Lastufka: When and why did you start making videos for YouTube?

Kevin Nalty: I've been making short comedy-like videos since high school and began posting them on the Web in December 2005. I didn't really begin using YouTube until late 2006 when I realized that none of the other sites had the size and regularity of YouTube.

Alan: How did you come to have your YouTube celebrity?

Kevin: Although most YouTube users are unfamiliar with the notion of a YouTube celebrity, there is a core group of maybe 20,000 people who are active viewers and commenters. Within this YouTube community, I began to be recognized after I made a video called *Viral Video Genius*. But it wasn't until *Farting in Public* that I began to develop a strong base of viewers and subscribers. Soon I realized that much of the fun of YouTube is making videos for this community and about the chronic YouTube users.

Alan: If you had one sentence to describe your YouTube channel, what would it be?

Kevin: Variety.

I fatigue quickly of viewing someone if they develop a predictable format. I hope that my videos vary enough for people, whether it's a vlog about the YouTube community, a short comedic skit, or an interesting moment in my life. Many seem to like seeing my kids in parody videos or candid moments where I'm out embarrassing myself.

Alan: How has your success on YouTube helped your career outside of YouTube?

Kevin: My day job hasn't changed much, but without my YouTube popularity I don't think I'd be as widely known in the online video community. Unlike other marketers, I have experience as a video creator. So, I have a unique ability to see the convergence of advertising and online video, and this has resulted in sponsor opportunities and speaking gigs.

Alan: Who is your target audience? And who is your actual audience?

Kevin: My blog targets video creators and advertising and marketing professionals, but my video audience is far broader. I have one of the most bifurcated audiences on YouTube, because it ranges from adolescents who share my sophomoric humor to parents who want to see my pathetic parenthood in action (probably so they can feel better about their own). When I look at the video demographics now visible on YouTube, I'll see a video about a snake in my pool drawing young guys and then a video about my kids stealing a van drawing the 30- to 50-year-old men and women equally.

Alan: In what ways do you interact with your audience?

Kevin: I probably interact more with my audience than most, but not as much as I'd like. Sometimes I'll ride the comment section of my most recent video for hours in a day, communicating with the hundreds of people who comment. I try to participate in collaboration videos whenever I can, and I've met many of the YouTube community members at events from California to New York City and from Georgia to London. My day job keeps me traveling, and that has also given me a chance to collaborate with people in various corners of the world.

Alan: Have your YouTube friends become your real life friends?

Kevin: I've developed dozens of contacts on YouTube that I'd consider friends—folks I'd invite to a party if they lived closer. We speak occasionally, email frequently, and get together when we can. I really like this group because they share my short attention span and creative fire, and their backgrounds are random. What I love most about YouTube Gatherings is that I can be talking to people with whom I'd otherwise have no contact—individuals from different regions, backgrounds, socioeconomic status, and job function. With maybe two or three exceptions, my geographic (in-the-flesh) friends are oblivious to the subculture of YouTube.

Alan: Has interacting with people on YouTube made you more confident while interacting with people face-to-face on a daily basis?

Kevin: Ironically, I actually think YouTube has made me a bit *more* introverted. I've had a taste of fame, and it can be fun and energizing in short bursts. But I love being anonymous and observing crowds. While it's fun to attend a YouTube Gathering and have people eager to meet me, I find it's even more rewarding to walk into a room and feel comfortable not being the center of attention.

Alan: How much of your day is spent on YouTube?

Kevin: Oh, this is hard to admit on paper. I probably spend from two to five hours a day on YouTube, an hour or so at night, and at least two in the morning. Sometimes I go "cold turkey" for a few days to gain perspective, but I really enjoy creating videos and learning what people like and don't like.

Alan: What advice do you have for up-and-coming YouTubers?

Kevin: Have fun and collaborate with the people on YouTube, and don't feel rushed about getting to some artificial milestone in the form of views or subscribers. Meet some people who arrive to YouTube about the time you do, and form some friendships. That's more exciting than getting featured, which results in lot of scrutiny. The most rewarding part of YouTube is when you develop a small, loyal following of people who share your interests or humor. So, focus on quality, not quantity. I'll take care of creating a video as frequently as you poop.

Alan: Could you repeat your success on YouTube if you started over today?

Kevin: Absolutely not. Just as I could never get into Georgetown today, I'd never be able to develop a robust subscriber base starting today. YouTube unfortunately provides a persistent advantage to those who developed followings early. If you look at the Most Subscribed list, you'll see many people who couldn't start over and are probably popular *because* they are popular. Some have learned to keep things fresh, and other amateurs slowly fade when the audience gets tired and moves onto the "star of the day."

Alan: You've participated in numerous collaboration videos; do you feel these types of videos are important?

Kevin: I was fascinated when a YouTube friend (Pipistrello) told me he meets with other YouTubers. I was a bit scared to get together with strangers from the Internet but was thrilled with the experience. It's so fun to be a brief actor in someone else's video or to join with other people to produce something. The viewers who are part of the YouTube community are amazed to see two people—who they watch separately—perform together. Collaborations are a faster way to develop an audience if it's done right, but more important, it's extremely satisfying from a creative perspective. I don't interact with many creative people at work, so my brain goes Technicolor watching how other people approach the conception and execution of a short video. I've learned so much by watching other people at work.

Interview with Liam Kyle Sullivan (liamkylesullivan)

YouTube: *www.youtube.com/user/liamkylesullivan* (URL 15.11)

Website: *http://liamshow.com/* (URL 15.12)

Liam Kyle Sullivan.

Liam has had videos with many millions of views on YouTube: *Shoes* has received 20 million views, *Let Me Borrow That Top* has received 10 million views, and *Muffins* has received 7 million views. As a result of his YouTube work, Liam has been a guest on the TV shows *Gilmore Girls*, *8 Simple Rules*, and *Alias*. And he's in the Weezer music video for "Pork and Beans." He's also opened for Margaret Cho on tour doing stand-up comedy.

He plays a female character, Kelly, in the majority of his videos.

Liam says this of his work: "In my comedy, I like playing characters who never let anyone grind them down and, of course, must eventually be carted away in straight-jackets."

Alan Lastufka: When and why did you start making videos for YouTube?

Liam Kyle Sullivan: I actually started making videos in 2004, before YouTube was invented. I had a live comedy show going in Los Angeles, the Liam Show, where I'd show my videos and perform characters and songs. I had just finished shooting the *Shoes* video when YouTube started getting popular, and a fan downloaded the video from my website and posted it on YouTube. And who knew—it got millions of views! I remember asking a friend, "What's YouTube?" I investigated, made my own You-Tube channel, posted my videos myself, put a whole "Kelly" album on iTunes, and here we are.

Alan: You've had numerous videos viewed tens of millions of times; what do you attribute to the success of these videos?

Liam: I guess people dig my sense of humor. I try not to think about why. If I over-analyze it, I get frozen.

Alan: If you had one sentence to describe your channel on YouTube, what would you say?

Liam: This channel is everything you've ever dreamed of and more!!!!!

Alan: How has your success on YouTube helped your career outside of YouTube?

Liam: I've been opening for Margaret Cho on her Beautiful Tour, and I've been performing at colleges across the country, which wouldn't have been possible before. I'm also free to develop new ideas, pitch ideas to studios (people with funding), and do a lot of paid writing.

Alan: Who is your target audience? And who is your actual audience?

Liam: I know a lot of young women watch my videos. But I've heard from all kinds of people—young, old, straight, gay, male, and female—who enjoy them. I try not to target anyone. I just try to do what I think is funny.

Alan: In what ways do you interact with your audience?

Liam: I've set up a fan page on my website where I post blogs and let people know what I'm up to. I don't normally respond to messages coming in from all the social networks.

Alan: Has your circle of friends from YouTube transferred to your circle of friends in real life?

Liam: No, I'm kind of a loner, actually. Socially stupid. A little boring, if truth be told. Most times I like to read and do the crossword for fun.

Alan: Has interacting with people on YouTube made you more confident while interacting with people face-to-face on a daily basis?

Liam: No. I'm mostly pretty comfortable with people I've known for a long time.

Alan: How much of your day is spent on YouTube?

Liam: I try to leave it alone; otherwise, I get all caught up in the numbers. It's easy to get obsessed and go ego-tripping. I have to remember that today's entertainers can be yesterday's news pretty quick.

Alan: What advice do you have for up-and-coming YouTubers?

Liam: Make videos you enjoy making. They might suck at first. Stick it out. It took me years to learn to write and shoot something good. Just keep going.

Alan: Do you feel you could repeat your success on YouTube if you started over today? Why or why not?

Liam: You start over every day, anyway, when you're creating stuff. Maybe I could repeat it; maybe I couldn't. I think I was successful on YouTube because I wasn't trying to be successful on YouTube. I was just trying to make something funny with my friends.

For additional interviews with YouTube rock stars, please visit the following site:

http://viralvideowannabe.com/interviews (URL 15.13)

Index

dekePOD™

Uncensored, Unregulated and Frankly Unwise

deke.oreilly.c

www.deke.com

Author Deke McClelland has been unleashed in dekePod, his new video podcast series on the topics of computer graphics, digital imaging, and the unending realm of creative manipulations. New podcasts are posted every two weeks. Drop by deke.oreilly.com to catch the latest show or to browse the archives for past favorites, including: "Don't Fear the LAB Mode," "Buy or Die: Photoshop CS4," and "101 Photoshop Tips in 5 Minutes."

Don't miss an episode when you subscribe to dekePode in iTunes

O'REILLY®